Flávia Gabriela Domingos Silva

The cartographic scale at your fingertips

Flávia Gabriela Domingos Silva

The cartographic scale at your fingertips

Contribution of tactile models to the construction of the notion of proportionality by visually impaired students

ScienciaScripts

Imprint

Cover image: www.ingimage.com

This book is a translation from the original published under ISBN 978-3-330-76513-9.

Publisher:
Sciencia Scripts
is a trademark of
Dodo Books Indian Ocean Ltd. and OmniScriptum S.R.L publishing group

120 High Road, East Finchley, London, N2 9ED, United Kingdom
Str. Armeneasca 28/1, office 1, Chisinau MD-2012, Republic of Moldova, Europe
Managing Directors: Ieva Konstantinova, Victoria Ursu
info@omniscriptum.com

Printed at: see last page
ISBN: 978-620-8-51592-8

I dedicate this book to my husband, Diego, to my parents, Divino and Djanira, to my brother Júnior, to my teacher and friend Dr Miriam Bueno. And in a special and affectionate way, I dedicate it to students with visual impairments and to the teachers who set out to teach and learn with these individuals.

SUMMARY

INTRODUCTION

The historical process of formal education for students with disabilities in Brazil was consolidated by the development of Special Education, which in the current context is structured as a support and training space for regular schools. This situation is considered to be the result of the struggle for the rights of people with disabilities and the development process of Special Education itself, which has historically had different phases.

Before discussing this type of education, it is necessary to define the terms that will be used, because as Sassaki (2005, p. 1) argues, terminology is "especially important when we approach subjects that are traditionally riddled with prejudices, stigmas and stereotypes, as is the case with disabilities". In this sense, the terminologies used in this work are based on this author's conceptions: people without disabilities; people with disabilities; visually impaired person (student); DV person (student); blind person (student); person (student) with low vision (SASSAKI, 2005).

The structuring of educational services for people with disabilities in Brazil began in the 19th century with the inauguration of the Imperial Institute for Blind Children in 1854. During this period, the pedagogical care of students with disabilities was carried out mainly in hospital environments based on medical-pedagogical processes. At the time, the paradigm of incapacity prevailed, in which "school education was not considered necessary, or even possible, especially for those with severe cognitive and/or sensory disabilities" (GLAT; FERNANDES, 2005, p. 2).

The development of pedagogy and learning psychology showed that people with disabilities could learn, driving the need to create appropriate learning conditions and spaces for them. At this time, special education was established through the multiplication of special schools, completely separate from mainstream education. The phase of segregation was established, which was reinforced by legislation, the educational structure and the socially shared mentality.

In the 1980s, demands against the exclusion of minorities gained momentum. And in Brazil, in the educational field, the integrationist movement emerged, which aimed to prepare students from special schools to join the ordinary classes of the regular education system. This perspective proposed the structuring of some specialised services, with the aim of linking special education to mainstream education (MAZZOTTA, 1993).

However, some authors argue that in the mainstreaming movement, the school retains its organisation and dynamics, so that the student, although included in mainstream education, remains excluded. "There is a consensus that integration presupposes 'tutored participation' in a structure with its own values, to which the 'integrated' student has to adapt" (RODRIGUES, 2006, p. 3).

Subsequently, the inclusive educational perspective took shape, which guides the discussions here. This concept gained strength with the Salamanca Declaration (1994), which proposes valuing diversity, equal opportunities in the educational process and offering quality education to all students. As a result, Special Education is re-dimensioned as a support for mainstream education, since students with disabilities are definitively included in this area of education.

In this way, mainstream education, as well as other areas of society, is part of the process of enabling the inclusion of people with disabilities. Emphasising what Mazzotta (1993) argues that the type, degree and consequences of a disability determine the re-dimensioning of the educational process, we highlight the demands arising from visual impairment.

Visual impairment is the absence or partial impediment of the sense of sight and extends from low vision to blindness (JORDÃO, 2011), and in relation to teaching, this disability is classified based on visual functionality. In view of the limitation or absence of sight, teaching students with visual impairment (VI) requires valuing multisensory experiences, especially those arising from the tactile sense, which is an important means of acquiring information for blind people and, in some cases, those with low vision, as it allows them to grasp different properties of objects and environments (BATISTA, 2005).

The development of spatial reasoning is school geography's major contribution to citizen education, and cartography is extremely important in making this process a reality. Therefore, because it enables the construction of skills that influence the subject's relationship with his/her place of living, the consolidation of cartographic knowledge must be realised with all students.

Considering the need to adapt the educational process for visually impaired students, it is essential to discuss the teaching of cartography in this context, bearing in mind that cartographic knowledge is essentially visual and has historically been constructed to be learnt through the sense of sight (ALMEIDA; NOGUEIRA, 2009).

Cartographic scale is one of the most important concepts in cartography. It is the "fraction that represents the relationship between a distance on the map (usually 1 cm) and the equivalent distance on the ground (X cm)" (LE SANN, 1984, p. 56). Understanding this concept is often difficult because it is a mathematical relationship between large areas in reality and extremely small representations. It is therefore argued that addressing non-quantifiable proportionality relationships present in students' living spaces helps them to understand the concept of cartographic scale, which is characterised by being a quantifiable proportionality relationship.

The notion of non-quantifiable proportionality is built on comparisons between objects, first and second order relationships and the definition of perceptual judgements such as greater than, equal to, less than (SPINILLO, 1994). When considered in the concept of cartographic scale, this proportionality relationship affects the variable size (distance) and is linked to the spatial dimension.

Based on the assumption that the understanding of the notion of proportion and the concept of cartographic scale is greatly influenced by the sense of sight, the following problem was developed: what is the contribution of tactile models to the development of the notion of proportionality, based on lived space, and the concept of cartographic scale by visually impaired students?

In the teaching-learning process of visually impaired students, the use of appropriate teaching resources is extremely important, as they make representations, information and data accessible, subsidising the process of building knowledge for these individuals. Therefore, the general objective was to evaluate the contribution of tactile models to understanding the notion of proportionality and cartographic scale by visually impaired students, as well as to developing their orientation and mobility.

This book is the result of the author's master's research, which was carried out using the Participatory Research methodology, defined as a qualitative approach consisting of three fundamental aspects: research, education and action. The relationship established between theory and practice in this methodological modality is characterised by dialectics. With regard to this discussion, Demo (1982) states that:

> This brings back the dialectical quality of the relationship between theory and practice. Both terms need and repel each other, in an identity of opposites. In other words, one does not exist without the other, but each has its own density, which makes a dynamic relationship possible. On the one hand, we have the absolutising tendency of theory. [...] Practice is a historical product, in other words, limited, relative, procedural and endless. Theory can be absolute, abstract, utopian, universal; practice, on the other hand, is relative, concrete, realised, particular (p. 62).

Based on the assumption that practice is indispensable to the applicability of theory, Participatory Research is considered to "intervene in real situations and not in laboratory situations. It involves working with real groups, with existing limitations and resources" (HAGUETTE, 2001, p. 151). The field of research was therefore defined as the Brazilian Centre for Rehabilitation and Support for the Visually Impaired (CEBRAV), which, among other services, caters for visually impaired students enrolled in the Goiás State Education Network.

In the search for a more in-depth understanding of the research subjects and better control of the proposed stages, the intervention in Participant Research takes place on a relatively restricted scale, and it is sometimes possible to generalise the results (HAGUETTE, 2001). We therefore chose to follow three students (two in secondary school and one in primary school) who attend the Maths Workshop[1] at CEBRAV.

Another aspect of Participatory Research is that it has a political meaning, since it directs its investigations towards marginalised and socially oppressed subjects, with the aim of understanding a certain existing problem and contributing to solving it, or at least minimising it (ITABORAHY, n.d.).

In this context, working with DV students based on Participant Research must consider the educational process of these subjects, which historically was based on the paradigm of incapacity and isolation from the regular system, and was later re-dimensioned by the integration movement and the inclusive concept, which in practice did not mean overcoming the precariousness of teaching conditions, the lack of accessible resources and practices, in other words, educational segregation.

Based on these aspects, the literature review was guided by the following themes. Geography teaching; school cartography; the concept of cartographic scale in school geography; the mathematical reasoning of proportionality; Piaget's spatial relations; the geographical concept of place and the theoretical-methodological foundations of tactile cartography.

Participatory Research also presupposes the joint participation of the researcher and those being researched, both of whom are subjects in the same process, in which knowledge is not unilateral, but is constructed through interrelationship and the exchange of experiences (BRANDÃO, 1999). Thus, the monitoring of the students began with the observation stage (first half of 2014).

According to Haguette (2001), the researcher must assume the position of a critical observer and active participant. To do so, they need to be familiar with the reality under investigation and have the sensitivity to understand the participants. In order to subsidise the process of designing and making teaching resources and planning and implementing the teaching sequence, a technical visit was made to the Teaching and Teaching Materials Laboratory (LEMAD) at the University of São Paulo and the Lara Mara Institute in São Paulo (SP). Braille and orientation and mobility classes were also held at CEBRAV.

The didactic sequence developed sought to identify the notion of proportion that the students have; to promote the students' contact with tactile representations, prioritising the improvement of the notion of proportionality and the construction of the concept of cartographic scale; to check whether the tactile representations developed actually contribute to understanding of this knowledge; and to assess the extent to which working with proportion, based on lived space, helps to develop an understanding of spatial organisation and orientation and mobility skills.

The aim of contributing to the construction of the notion of proportionality and the concept of cartographic scale

[1] CEBRAV's Pedagogical Support Unit offers workshops, which are held weekly in the off-shift, based on the interdisciplinary axes: "interaction, communication and language; sensory-motor skills; autonomy and social competence; citizenship and knowledge related to the areas of Portuguese, Maths, Chemistry, Art, Culture and Sport" (CEBRAV website, 2011).

by DV students and, to some extent, improving their orientation and mobility skills, is related to the central proposal of Participant Research, which advocates improving the quality of life of the participants, as well as expanding their capacities for analysis, critical awareness and problem solving.

Although the development of accessible teaching resources and appropriate teaching methodologies are extremely important for teaching students with disabilities, the approach to these issues in undergraduate courses is not satisfactory (FREITAS; VENTORINE, 2011), and Geography courses, as well as specifically the teaching of Cartography to visually impaired students, are part of this context.

The precariousness of initial training in the education of people with disabilities means that improving the quality of this teaching depends almost exclusively on continuing training, as Carmo (2011, p. 254) shows:

> The skills to respond to the needs posed by this new educational reality, which is inclusion, should be provided for in initial teacher training, but most courses don't include enough discussion [...] That's why continuing training courses are the only opportunities to have access to these discussions.

The obstacles in teacher training processes contribute to the precariousness of the education offered by mainstream schools to visually impaired students. These institutions are generally characterised by a lack of physical structure, specific teaching resources and professionals who are effectively trained to lead the teaching-learning process for these individuals. However, there is an urgent need to change this situation, given that mainstream educational systems, influenced by the inclusive perspective, are increasingly responsible for the formal education of students with disabilities.

In this context, it is hoped that the discussions and results achieved in this research can provide theoretical and methodological assistance to geography teachers who deal with visual impairment in their teaching activities, contributing, even if only minimally, to the consolidation of truly inclusive teaching.

Chapter I, entitled "The education of visually impaired students in the Goiás State Education Network", discusses the formal education of these individuals, considering the process of structuring Special Education in the Goiás State Education Network. Topic 1.1 presents the characteristics and implications of visual impairment and the educational specificities of VI students, as well as the specialised educational care offered in the Goiás State Education Network. Topic 1.2 emphasises the work carried out by the Brazilian Centre for Rehabilitation and Support for the Visually Impaired (CEBRAV) and finally, subtopic 1.2.1 discusses the supervising teacher's conceptions of the issues raised.

Chapter II, "The place where I live fits in my hand: School Cartography, Tactile Cartography and the concept of cartographic scale" began with a discussion of the importance of School Cartography in Geography teaching, emphasising the need to articulate the knowledge and practices of Cartesian cartography with those of non-Cartesian cartographies. Later, in topic 2.1, the concept of cartographic scale was discussed, especially in the context of school cartography. Topic 2.2 discusses the construction of the notion of proportionality from a mathematical point of view and presents a reflection on Jean Piaget's spatial relations. Finally, topic 2.3 discusses the theoretical and methodological foundations of Tactile Cartography. Subtopic 2.3.1 presents the methodological stages of developing and making tactile representations, and subtopic 2.3.2 discusses the articulation of Tactile Cartography with Geography teaching.

Entitled "Methodological stages: guiding and being guided in the construction of the path travelled", Chapter III began by describing the structuring of the research, explaining the choice of CEBRAV as the field of research, the observation stage and the selection of spaces for representation. Topic 3.1 deals with the characteristics of the participants, topic 3.2 describes the whole process of designing and making the tactile teaching resources, with subtopic 3.2.1 discussing the evaluation of the materials, highlighting the importance of the Sensory Notebook, subtopic 3.2.2 presents the construction of the tactile teaching resources, which was based on the foundations of Tactile Cartography and subtopic 3.2.3 deals with the evaluation of the tactile cartographic representations. Finally, topic 3.3 discusses the process of preparing the didactic sequence and subtopic 3.3.1 presents the didactic sequence used in this research.

Chapter IV, entitled "Didactic proposal for the construction of the notion of proportion and cartographic scale with visually impaired students" describes the activities carried out with the research subjects in topics 4.1, 4.2 and 4.3 respectively. And topic 4.4 analyses and discusses the results obtained from the aforementioned activities.

Finally, the Final Considerations take up the main themes and reflections made throughout the work, emphasising the analysis of the results of the activity carried out with the research subjects.

CHAPTER 1

THE EDUCATION OF VISUALLY IMPAIRED STUDENTS IN THE STATE EDUCATION NETWORK OF GOIÁS

Special education in Goiás is structured in conjunction with the national situation, which in turn is shaped by perspectives resulting from the struggle for the rights of people with disabilities on a global scale. As well as state laws and documents, which despite being based on general principles, take into account the specificities of the local reality.

In the Brazilian context, the development of Special Needs Education has undergone a re-signification of its conceptions and characteristics, as well as state actions. It originated based on medical models and welfare actions, and as it acquired an effectively educational character, it established itself as a parallel and isolated system to Regular Education, later being re-dimensioned from the perspective of inclusion.

Mazzotta (2005) argues that the process of structuring Brazilian special education can be divided into two periods. The first, from 1854 to 1956, is characterised by isolated public and private initiatives; and the second, from 1957 to 1993, is marked by the emergence of official nationwide actions, which continue to be developed to this day.

The institutionalisation of educational care for people with disabilities in Brazil took place mainly at the end of the 1950s, and in the two decades that followed it was strongly influenced by the medical-psychological perspective (MAZZOTTA, 1996). It was at this juncture that Special Education began in Goiás, considering that in 1953, through Law No. 926, the Pestalozzi Institute of Goiânia was created. This institute remained the only public institution offering services to people with disabilities in the state until 1973.

In the national context, the 1970s marked the emergence of the discussion on the education of people with disabilities in the Law of Guidelines and Bases of National Education. This approach, which was neglected in the 1961 LDB (Law no. 4,024), was present, albeit minimally, in the document approved in 1971 (Law no. 5,692). Thus, this legislation:

> It defined the need for special treatment for students with disabilities (physical and mental), those who were considerably behind in age and the gifted. By not organising an education system to meet specific needs, [the LDB/71] reinforced the referral of these students to special education institutions (SILVA, 2011, p. 18).

In 1973, by Decree 72.425, the Ministry of Education (MEC) inaugurated the National Centre for Special Education (CENESP), responsible for disseminating, organising and administering Special Education in Brazil. That same year in Goiás, the first body responsible for the education of people with disabilities in the state was created, the Special Education Section, linked to the Supplementary Education Department of the State Department of Education. In 1974, this section was renamed the Special Education Division and was linked to the 1st Grade Education Unit.

Teacher training courses and the creation of scholarship programmes and special classes in regular education[2] were some of the activities carried out by the Special Education Division. Following the national context, the actions aimed at educating people with disabilities in Goiás were based on the concept of integration.

The mainstreaming movement proposed a relative rapprochement between Special Education and Regular

[2] According to the National Guidelines for Special Education in Basic Education (BRASIL, 2001) "Special Class is a room in a regular school, with adequate physical space and modulation. In this type of classroom, the special education teacher uses specialised teaching methods, techniques and procedures and teaching resources and, when necessary, specific teaching equipment and materials, according to the grade/cycle/stage of basic education that the student has access to the common national curriculum" (p. 53).

Education, since the ability of disabled students to adapt to the ordinary school would define whether or not their educational process would take place in that system. In this way, various services are offered to help integrate students, such as itinerant teaching and resource rooms; special classes in ordinary schools; special schools and hospital or home teaching (MAZZOTTA, 1993).

Although the integrationist perspective is considered by some scholars to be a possibility for the democratisation and humanisation of the regular education system (SANTOS, 2002), criticism of this concept is based on the fact that the student is solely responsible for their entry and permanence in the regular school, and if they fail to adapt to the school reality, they have to return to Special Education (RODRIGUES, 2006).

In 1978, the MEC proposed the Priority Project for Reformulating Curricula for Special Education, which, according to Glat and Fernandes (2005), established that disabled students could be cared for in clinics, rehabilitation centres or regular schools. This document reinforced the integrationist conception, as it presented the possibility of access to regular education, but did not consider the reformulation of this system.

In 1980, Law No. 8,780 was enacted, which defined measures for the Goiás State Education Network. Article 27 of this document deals with special educational treatment. However, according to Almeida (2003), this legislation does not mention people with physical and sensory disabilities, which minimised the contribution of this law to the special educational services offered at the time.

The structuring of Special Education in Goiás made relative progress in 1982, when Ordinance 1.674 abolished the Special Education Division and created the Special Education Unit (UEE), linked to the Superintendence of Educational Affairs of the State Department of Education (SEE). This new body:

> It greatly increased the number of special classes in regular education, covering several municipalities, and began to systematise early care, pre-school, primary, secondary and secondary education, as well as the professionalisation of students in all areas of disability. It also began to cater for the gifted (MAKHOUL; et. al., 2010, p. 7).

The main aim of these actions was to make the state's special education services effectively educational, definitively abandoning the clinical model[3] that had existed until then. While at national level, in 1986, by Decree 93.613, CENESP was replaced by the Special Education Secretariat (SESPE), in Goiás, a year later, by Law 10.160, the extinction of the UEE was legitimised and the Superintendence of Special Education (SUPEE) was created, which was directly linked to the State Education Secretariat.

In order to direct and coordinate special education throughout the state, SUPEE established partnerships with other education superintendencies, regional police stations, city halls and universities (ALMEIDA, 2003). However, despite the expansion and structuring of the educational services offered to people with disabilities, Special Education in Goiás did not overcome the perspective of integration, i.e. the responsibility for adaptation remained with the students and not with the education systems.

In 1990, SESPE was extinguished and its attributions were redirected to the National Secretariat for Basic Education (SENE), and this approach, albeit in administrative terms, revealed "historically, a timid attempt to overcome the old dichotomy between regular and specialised education in national education" (BEZERRA; ARAUJO, 2014, p. 104). However, in 1992, Special Education in Brazil was once again represented by a specific secretariat and through Law No. 8,490, the Special Education Secretariat was re-inaugurated, now with the acronym SEESP.

---

[3] In the clinical model, disability is conceived as a disease and care for people with disabilities, including educational care, is carried out for therapeutic purposes (MAKHOUL; et.al., 2010).

In 1994, the Salamanca Declaration, which is based on the principles of inclusion and education for all, became widespread in education debates. This document establishes that students with special educational needs should have access to mainstream schools and that the responsibility for adaptation is centred on the education system and no longer on the individual. This position makes special education a support, training and resource centre for mainstream schools (UNESCO, 1994).

This declaration discusses, among other things, the target audience for Inclusive Education, thus defining that the term Special Educational Needs is not restricted to students with disabilities, but refers to all those who for different reasons, social, economic, cultural, present difficulties at school.

Another aspect addressed in the Salamanca Declaration (1994) is the importance of reflecting on and materialising a flexible curriculum. It is understood that the open curriculum, i.e. one that presents learning as the centrepiece and takes into account the local context and the individual differences of the students, is the model that is most coherent with the Inclusive Education proposal (Carvalho; Carvalho; Custódio, 2010).

Curricular versatility, however, does not imply the construction of different curricula. All pupils, regardless of their cognitive and physical potential, have the right to access the curriculum, with the contents, methodologies and teaching resources being the elements that can be adapted to suit the various specificities.

Although the proposals of the Salamanca Declaration (1994) were far-reaching, the guidelines and actions aimed at Special Education in Goiás did not disregard the integrationist concept. In 1995, the Special Education in Goiás document was published, defining the concept and objective of this type of education. Thus, special education is understood as a process that permeates the various levels and grades of education:

> It has the same aims and objectives as mainstream education and aims to provide students with special educational needs with specialised care throughout their development and learning process, so that their potential is worked to the maximum, according to their own pace and potential, seeking to provide them with the necessary conditions for the full exercise of their basic rights and their effective social and occupational integration into society (ALMEIDA, 2003, p. 22).

Also according to this document, it is proposed that care for people with special educational needs should take place in various spaces[4] defined according to the needs of each student. Almeida (2003) argues that the determinations of the Special Education in Goiás Document (1995) did not promote the interrelationship between Special Education and Regular Education, because:

> Of the nine types of special education offered, five were located in regular education, although special education was responsible for managing and monitoring the types of education offered, training its teachers and referring previously diagnosed students (p. 24).

In the national context, the 1996 Law of Guidelines and Bases for National Education (Law No. 9.394), unlike the previous ones, expanded the proposals regarding Special Education in Brazil, presenting a chapter for the discussion of this education, although it did not advance in terms of the inclusive concept. Article 58 defines Special Education as a "type of school education offered, *preferably* in the regular school system, to students with special needs" (BRASIL, 1996) (my emphasis).

It should be noted that in addressing the need for specialised teachers, the LDB of 1996 makes explicit the educational perspective that guides the determinations about Special Education in the document. Article 59, item III states

[4] The various educational spaces correspond to: special school, special class, ordinary class, integrative class, resource room, ordinary class with specialised support, pedagogical workshop, essential stimulation room and hospital and home care.

that education systems must ensure "teachers with appropriate specialisation at secondary or higher education level for specialised care, as well as regular education teachers trained to *integrate* these students into ordinary classes" (my emphasis).

A positive element of the LDBN (1996) is that it addresses aspects that are essential for teaching students with disabilities, such as the need for different educational spaces, the adaptation of curricula, methods, techniques and educational resources, flexibility in the time it takes to complete primary school and equal access to the benefits of social programmes.

Almeida (2003) states that although there have been advances in organisational and legal issues at national and state level, Special Education in Goiás did not change significantly between 1987 and 1998[5] , as SUPEE's working philosophy remained integration, even though the discourse on inclusion was gaining more and more strength.

In 1999, Goiás' special education policy began to be based on an inclusive concept, and important changes were made to the structure of this type of education, including the abolition of the Screening and Diagnosis Department; a partnership with the Superintendence of Vocational Education and the State Department of Citizenship and Labour; the restructuring of special schools; the formation of the Inclusion Support Sector[6] ; and the transfer of disabled students from special schools to regular schools. Also that year, the Superintendence of Special Education changed its acronym to SUEE.

The process of reformulation that began in 1999 continued in 2000, when the Education for Diversity in an Inclusive Perspective Programme - Peedi - was launched. With the aim of implementing an inclusive educational policy in Goiás, it proposed reformulating political-pedagogical projects, adapting the physical structures of schools and training human resources (MAKHOUL; et. al., 2010).

Based on Vygotsky's socio-cultural theory, ten projects that made up Peedi were developed and launched: Hoje Project, Caminhar Juntos Project, Prevenir Project, Communication Project, Depende de Nós Project, Awakening Project, Creative Space Project, Re-Fazer Project, Reference Units Project and Inclusive School Project. The actions of this programme were planned to take place over four years: 1999 - awareness raising; 2000 - implementation; 2001 - expansion and 2002 - consolidation.

In 2002, the Superintendence of Special Education (SUEE) offered courses in Inclusive Education to professionals from all over the state, as well as forming the Inclusion Support Network (REAI), made up of professionals from the Inclusion Support Sector, resource teachers and support teachers, whose aim was to advise school units.

Efforts to disseminate the inclusion policy in the state of Goiás were intensified in 2004, when the Itinerant Laboratories began to take place in the regional education sub-secretariats and in the municipalities that joined the programme. And the implementation of the Alternative Room Project, which aimed to give continuity to the teaching process for "young people and adults with special needs, who were left over from the special rooms" (MAKHOUL, 2010, p. 27).

---

[5] In 1998, through a partnership between the Brazilian Association of Educators of the Visually Impaired and the State Department of Education, the Pedagogical Support Centre for Visually Impaired People - CAP/GO - was inaugurated.

[6] The Inclusion Support Sector - SAI is a "team of professionals assigned to the Reference Unit Project, with the aim of subsidising teachers in inclusive schools and special schools in terms of strategies, teaching resources and curriculum adaptation, guiding and monitoring the development of students with special needs" (MAKHOUL; et. al., 2010, p. 20).

Based on the need for specific care for people with hearing impairments, the Centre for Training Education Professionals and Care for People with Deafblindness (CAS) was built in 2005. In 2006, CEE Resolution 07 was approved, which significantly influenced Special Education in Goiás.

This Resolution reinforces the idea that the regular education system is the locus of development for Special Education, so that students with special educational needs[7] should be catered for in ordinary classes and it is up to the school to adapt to the demands of these individuals. Even so, this document does not rule out the contribution of special classes in regular schools[8], which should be of a transitional nature, and of special schools, stating in article 17 that:

> Students with special educational needs, who require individualised attention in the activities of independent and social life, resources, aids and intense and continuous support, as well as curricular adaptations so significant that the regular school has not yet been able to provide them, must be catered for, on an exceptional basis, in public and private special schools (GOIÁS, 2006).

In order to broaden the possibilities for including students with SEN in mainstream schools, CEE Resolution 07 sets out the necessary aspects for school adaptation. As such, it stipulates that regular education institutions must provide accessibility in buildings; qualified or specialised teachers and technical-pedagogical staff; specialised teaching support; a reduction in the number of students per class; complementary and supplementary specialised educational care; flexibility and curriculum adaptation; curriculum enrichment and acceleration projects for the gifted; the provision of bilingual education, Portuguese language and Libras and the provision of the Braille system (GOIÁS, 2006).

The resolution's approach to the curriculum points out that it should take into account the teaching and learning characteristics of students with disabilities, special educational needs and high abilities, in order to reshape the school's pedagogical practice. The discussion of the curriculum is linked, in article 15, item VII, to the flexible timing of the school year, which should be used at "any stage of the schooling flow, to cater for students with special educational needs, so that they can complete the planned curriculum in more time, in an attempt to avoid a large age/grade gap" (GOIÁS, 2006).

Specialised pedagogical services are also covered in CEE Resolution 07. Of particular note is the work of the support teacher, who according to Tartuci, Cardoso and Freitas (2013) is responsible for assisting students with disabilities who, in order to develop curricular activities, need intense and continuous monitoring. As these students attend regular classrooms, the work of the support teacher subsidises the work of the regular classroom teacher.

In addition, there is the resource teacher who is responsible for subsidising, complementing and guiding pedagogical activities. These professionals work in resource rooms, which must contain specific equipment, furniture and materials suited to the students' educational needs (GOIÁS, 2006). Other professionals such as a Brazilian Sign Language (LIBRAS) instructor, a Brazilian Sign Language (LIBRAS) interpreter, a Braille instructor and an administrative hygiene support professional are also mentioned in the Resolution as collaborators in the implementation of specialised educational services.

---

[7] Article 4 of CEE Resolution no. 07 defines students with special educational needs as those with: **I** - limitations in the development process and/or marked learning difficulties in curricular activities, understood as: (a) those not linked to a specific organic cause; (b) those related to conditions, dysfunctions, limitations or disabilities; (c) those arising from neurological and psychiatric syndromes and severe psychological conditions; **II** - communication and signalling difficulties, differentiated from other students, for whom differentiated forms of teaching and curricular adaptations should be adopted, using applicable language and codes; **III** - high abilities/gifted (GOIÁS, 2006).

[8] It should be noted that, according to CEE Resolution 07, students enrolled in special classes must have the same disability or syndrome and that the teachers in these classes need to be specialised or trained in the area of manifestation of the students' disabilities or syndromes (GOIÁS, 2006).

Also in 2006, the High Ability/Gifted Activities Centre (NAAH/S) and the Diversity Care Centre were set up, in addition to the implementation of the Inclusion Reference Schools model, which were restructured by the government, especially in terms of services and human resources, and received students from special schools, offering them specialised care (TARTUCI; CARDOSO; FREITAS, 2013).

In 2007, the Convention on the Rights of Persons with Disabilities was drawn up, influencing the structuring of special needs education at international, national and regional level. Based on the inclusive perspective, where education is established as a fundamental right, this document defines, in its article 24, that States Parties shall ensure that:

> a. People with disabilities should not be excluded from the general education system on the grounds of disability, and children with disabilities should not be excluded from free and compulsory primary or secondary education on the grounds of disability;
> b. People with disabilities can access inclusive, quality and free primary education and secondary education on equal terms with other people in the community in which they live;
> c. Reasonable adaptations according to individual needs are provided;
> d. People with disabilities receive the necessary support within the general education system to facilitate their effective education;
> e. Individualised and effective support measures are adopted in environments that maximise academic and social development, in line with the goal of full inclusion (UNESCO, 2007).

In the state context, in 2007, SUEE established a partnership with the Chemistry Institute of the Federal University of Goiás, forming the Goiás Network for Interdisciplinary Research in Special Education, which aimed to articulate teaching and research, promoting in-depth studies on the inclusion of education professionals. With regard to the process of professional training, the Document of the Convention on the Rights of Persons with Disabilities (2007) argues that:

> It will incorporate disability awareness and the use of appropriate modes, means and formats of augmentative and alternative communication, and pedagogical techniques and materials, as supports for people with disabilities (UNESCO, 2007, Article 24).

In 2008, SUEE became the Coordination of Special Education (COEE) and was linked to the Superintendence of Basic Education. That year, a number of studies were launched, including the Diagnostic Study, which was carried out in special schools with the aim of verifying the work carried out in these institutions. The results obtained in this study supported the construction/consolidation of the Specialised Educational Assistance Centres (CAEE).

The process of restructuring special schools into Specialised Educational Care Centres continued in 2009, when work also began to orientate regular schools so that they could contribute to providing specialised care for students with SEN. According to the Operational Guidelines of the Special Education Department for the Inclusion Support Network (s/d), specialised educational care is defined as:

> Set of activities, accessibility and pedagogical resources organised institutionally, provided in a complementary or supplementary way to the education of students in regular education. These activities are offered in multifunctional resource rooms [in mainstream schools] equipped with equipment, furniture and teaching and pedagogical materials for the provision of AEE. [And also] in CAEEs and in the state network's school units during the learning extension shift. Special education students will be assisted in the ESA by the ESA teacher, who will have to be grouped according to specific needs, and may be assisted individually and/or in groups of up to 6 students. (GOIÁS, s/d, p. 3).

According to this document, the AEE teacher must have a degree in Pedagogy or a degree in non-critical areas; be a permanent employee of the State Education Department; be a specialist in the area of Special Education, presenting certificates of training courses with a minimum workload of 120 hours in the area of Special Education, mainly in Libras, Braille, Soroban, among others (GOIÁS, s/d).

In 2011, the Ministry of Education, by Decree No. 7,480, determined the extinction of the Secretariat for Special Education (SEESP), transferring its attributions to the Secretariat for Continuing Education, Literacy, Diversity and Inclusion (SECADI). It is noteworthy that matters relating to Special and Inclusive Education in Brazil are linked to

SECADI to this day.

That same year, another National Decree (No. 7.611) influenced the structuring and development of Special Education from an inclusive perspective in Brazil. It is noteworthy that this document changed the nomenclature from resource room to multifunctional resource room, according to Tartuci, Cardoso and Freitas (2013):

> It concerns the distinction that used to be made in relation to the services provided, which in the first case [resource room] is defined by categories of disability and in relation to the second [multifunctional resource room] there is no longer this distinction it is to cater for all students with special educational needs - students with disabilities, global development disorders, high abilities or giftedness (p. 8).

Article 5 of the aforementioned decree deals with technical and financial support for Special Education, determining that in order to expand specialised educational care for students with disabilities, global development disorders and high abilities or giftedness enrolled in the regular public education network, the Federal Government is responsible for providing financial support to the public education systems of the states, municipalities and the Federal District, and to non-profit community, confessional or philanthropic institutions.

Still in relation to the financial aspect, article 9 of Decree 7.611 states that "for the purposes of distributing FUNDEB resources, dual enrolment of students in regular public education who receive specialised educational care will be allowed" (BRASIL, 2011).

Although Special Education in Goiás, as in the national context, has been constituted as a parallel and isolated system to regular education and has remained, for a long time, based on an integrationist conception, we realise that the legislation, albeit with some weaknesses, has pointed towards reflection and consolidation of a more inclusive education.

What stands out, therefore, is the increasingly conscious defence of the need to value diversity and build a mainstream school that enables equal opportunities in the teaching process. To this end, there is a clear need to adapt physical spaces, make curricula more flexible, develop accessible teaching methods and resources and provide specialised teacher training. These factors have been addressed and discussed over time in legislation on the education of students with disabilities.

However, it is essential to consider that there is a distance between legal determinations and school practice, and that sometimes the dynamics of the latter are not consistent with the organisation established by the former. Thus, considering that more precise statements require an empirical analysis of the education offered to students with disabilities in the State Education Network, we can infer that in terms of practice, the consolidation of inclusion in Goiás is further away than the official documents would have us believe.

1.1 Visual impairment and its educational specificities in the context of the Goiás State Education Network

Amongst the students served by Special Education and who are therefore directly involved in proposals and projects based on educational inclusion, those with visual impairments stand out. In this sense, the characteristics and implications of this disability and the educational demands arising from it will be discussed, as well as the services offered to visually impaired students in the state of Goiás, emphasising the work carried out by the Brazilian Centre for Rehabilitation and Support for the Visually Impaired (CEBRAV).

Visual impairment is a sensory limitation characterised by a reduction or absence of the ability to see, ranging from low vision at different levels to blindness. The World Health Organisation defines this disability based on the visual field[9] and visual acuity[10] , and this classification is carried out using the Snellen Decimal Optometric Scale (Figure1).

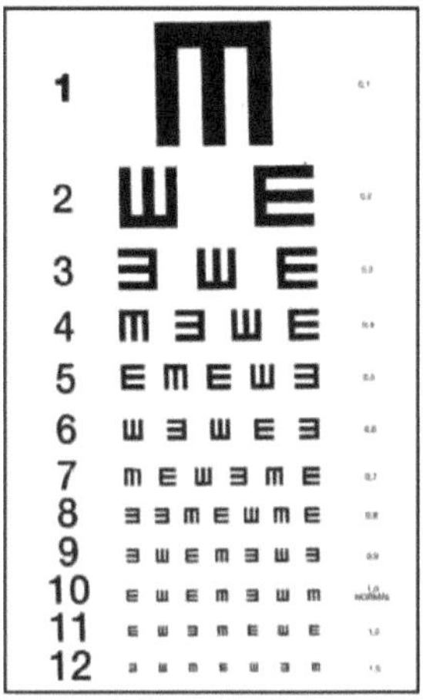

Figure 1- Snellen's Optometric Decimal Scale
Source: CMDV - Portal do Deficiente Visual - www.cmdv.com.br

According to Fonseca (1987), the top symbol of the Snellen Scale, used in visual acuity tests, measures 3.5 inches and can be seen by a normal eye at 200 feet (60 metres). This defines the occurrence of blindness:

> When the vision in the better eye is < 20/200, i.e. if you can see at 20 feet (6 metres) what a normal person can see at 200 feet (60 metres), or if the visual field is < 20 degrees in the better eye, even if the visual acuity in this narrow field of vision is greater than 20/200 (MIRANDA; PAULA, 2010, p. 6-7).

Low vision, in turn, is determined when, after correction or treatment, the individual has 20/70 vision in the better eye and is able to use this sense to perform tasks (MASINI, 1993). The way and conditions in which people with low vision see are quite diverse and visual residue is influenced by the type and extent of the pathology, age of vision loss, opportunities for visual stimulation and psychological conditions.

Considering that "the assessment of visual acuity alone is not a determining factor in the detection of visual impairment" (SEESP/MEC, 2006, p. 39), clinical classification is insufficient for educational work with VI students, and functional vision assessment is used for this purpose. The differences between these classifications are explained in Table 1:

CLINICAL CLASSIFICATION	EDUCATIONAL CLASSIFICATION
• Medical diagnosis - based on visual acuity. • Emphasis on what you see. • Legal purpose, economic and statistics. • Static result in conditions special distance and lighting. • Quantitative (numerical) data	• Educational diagnosis - based on Visual Efficiency. • Emphasis on how you see. • Practical and functional purpose in terms of OM performance in ADL and school tasks • Dynamic results in practical life conditions. • Qualitative data

Chart 1- Medical and educational classification: parallel and intersection Source: Masi (et. al., 2002, p. 27)

[9] Visual field "is all visual information received simultaneously (180°)" (MIRANDA; PAULA, 2010, p. 7).
[10] Visual acuity "refers to the measure of the ability to clearly distinguish the smallest details" (MIRANDA; PAULA, 2010, p. 8).

According to the educational classification, a student with low vision is one whose teaching-learning process is developed through visual perception, articulated with the other senses. Blind students, on the other hand, have anything from a slight perception of light projection to a total absence of the sense of sight, and the teaching process takes place by valuing the other senses and using the Braille system as the main means of written communication (SEESP/MEC, 2006).

Because they address different aspects, which are sometimes complementary, the integrated analysis of both classifications makes it possible to learn more specifically about the type and degree of visual impairment, as well as its implications for the construction of orientation and mobility skills, activities of daily living and the educational performance of the VI student, making it interesting for education systems to consider both types of assessment.

The sense of sight is extremely important in the subject's relationship with the world and its limitation or absence consequently interferes with various personal and social aspects of the individual:

> It is well known that vision transmits information [...] quickly and accurately, anticipates and coordinates movements and actions and accounts for 80 per cent of the individual's relationship with the world. Therefore, the implications of visual impairment for an individual's integration are many and significant, since the absence of sight impairs their understanding of the world and interferes with the quality of their exchange with the environment, often causing deprivation of experiences, limitation of movement and interference in spatial orientation (SEESP/MEC, 2006, p. 88-89).

The influence of visual impairment on the acquisition of knowledge and on understanding and adapting to the environment begins with the specificities caused to the process of sensorimotor development, since this relates, among other things, to "the action of the visual system in charge of mobilising the head and the whole body with eye movements through visual search reactions to light, brightness and moving objects" (BRUNO, 1993, p. 14).

Visual exploration of the environment helps to strengthen the neck muscles through tone, which enables cephalic control and keeping the head in a midline position. Thus, according to Bruno (1993):

> Children with severe visual loss perform little cephalic rotation for auditory search because they don't focus on the sound source in the early stages of development. [...] The auditory reaction is not automatic, it is slower, it depends on the distance and position of the stimulus in space (p. 15).

In addition, visually impaired children tend to reject postural changes and are insecure about moving and being moved, which interferes with their body awareness and their body's displacement in space, consequently slowing down their sensorimotor development. The lack of visual-tactile-kinesthetic coordination, made possible by the association between looking and moving the hands, hinders the execution of the grasping movement, as explained by Bruno (1993):

> As visually impaired children can't see the object, they don't anticipate the grasping behaviour. We believe that this is the crux of these children's sensorimotor development. In the absence of sight, the coordination of these schemes will be done by integrating the tactile-kinesthetic-auditory schemes through the sucking and grasping behaviours (p. 17).

Considering the interference of visual impairment on sensory-motor experiences, the evolution of the mechanisms for adapting to the environment are also considerably impaired, which directly influences the construction of the notion of space. This construction is established through the possibilities of displacement and orientation with one's own body.

In children with visual impairment, the notion of space occurs through the motivation of touch, the search for sound and the position and relationship of the body and objects in space, which makes it essential for them to have "the opportunity to experience experiences functionally, being able to do things and make discoveries with their whole body" (BRUNO, 1993, p. 18).

The construction of the meaning system, which corresponds to "the child's ability to act and perceive their own actions and those of others in order to be able to imitate them" (BRUNO, 1993, p. 20) is also hampered by the absence or limitation of the visual sense, above all due to the lack of direct observation of people, objects, attitudes and situations and the fragility of the information obtained by the tactile sense, which most of the time is not yet trained in the purposeful and potential search for data.

The help of other people in the construction of the meaning system of children with VI is essential, because it is through the expansion of their action on the environment, the experience of significant multisensory experiences and, above all, interaction and relationships with people, that the structuring of this system is made possible in these individuals.

The limitations arising from visual impairment, which are present from the very beginning of development, reinforce the need to detect the type and degree of impairment as soon as possible, as well as to carry out activities for the early stimulation of children with visual impairment. With regard to this discussion, the document Knowledge and Practices of Inclusion: developing competences for meeting the special educational needs of blind and low-vision students (SEESP/MEC, 2006) states that:

> The early detection of any of these problems can be a decisive factor in the child's overall development, as long as the right stimulation conditions are provided for their maturing needs, favouring the maximum development of their potential and minimising the limitations imposed by visual impairment (p. 18).

The specificities of how visually impaired children, adolescents and adults obtain knowledge, which is achieved through their remaining senses and visual residue, impose specific demands on educational institutions (schools and care centres). This makes it necessary to adopt appropriate practices, methodologies and teaching resources so that the teaching-learning process for visually impaired students can be consolidated.

The first attitude of educational institutions towards visually impaired students, which will influence other aspects of the teaching process, is to understand that the particularities of this individual, resulting from their disability, do not make them better or worse than others. In such a way that "realising this individual in their entirety, as a subject who has fears, dreams, difficulties and talents, helps to overcome attitudes of compassion, pity or overprotection" (SILVA, 2011, p. 31).

Given the importance of the remaining senses (touch, hearing, smell and taste) in the lives of people with visual impairments, activities that enable the development and enhancement of these senses, based on meaningful multisensory experiences, should be part of the teaching-learning process for students with visual impairments.

Among the remaining senses is the importance of touch, which, through training, is classified as active and is characterised by enabling the purposeful search for information and its comprehension (SILVA, 2011). This sensory system intensifies the acquisition of knowledge about environments and objects, as well as enabling literacy in the Braille system[11], making it indispensable to the teaching process of visually impaired students.

According to the document Knowledge and Practices of Inclusion: Developing competences to meet the special educational needs of blind and low-vision students (SEESP/MEC, 2006), schools must:

> Offer the visually impaired person as wide a variety of materials as possible, such as: different types of paper,

[11] "The Braille System, created by Louis Braille in France in 1825, is universally known as the code or means of reading and writing for blind people. It is based on a combination of 63 dots representing the letters of the alphabet, numbers and other graphic symbols. The combination of dots is obtained by arranging six basic dots, organised spatially in two vertical columns with three dots to the right and three to the left of a basic cell called the Braille cell" (MIRANDA; PAULA, 2010, p. 12).

> fabric, wood, leather, carpet samples, yarn, plastic, sandpaper, etc. With these materials, ask them to discriminate between thickness, size and texture: thick, small, big, smooth, rough, etc. With these materials, ask them to discriminate between thickness, size and texture: thick, thin, small, large, smooth, rough, soft, rough, etc (p. 47).

Hearing also makes a significant contribution to the learning of visually impaired students, both by improving the exploration/perception process, by associating with other senses, and by providing information about environments, objects and events that are not accessible by touch, through the speech of other people. However, just like the tactile sense, hearing needs to be developed. Silva (2011) states that:

> It should first be emphasised that, given the impossibility of controlling the sounds emitted by a given environment, blind children need to develop selective perception. Understanding the relationship between the sound object and the sound heard is a very complex exercise, especially when it is not possible to establish physical contact with the object. It is therefore essential that the family and school encourage the development of the sense of hearing from the earliest years (p. 32).

Teaching students with low vision also requires the use of appropriate resources, including materials specifically designed to improve the visual capacity of students with low vision, called optical and non-optical resources, and those relating to the teaching process itself, for pedagogical purposes.

Optical resources are characterised by high magnification lenses designed for visual correction, and near vision is aided by the use of bifocal or monofocal glasses, telemicroscopic systems and manual and support magnifiers. For long-distance vision, teleloupes are generally used, which allow the object to be seen to be magnified and brought closer (SEESP/MEC, 2006).

Non-optical resources include changes to the environment, enlarged type, yellow acetate, inclined planes, accessories (4B or 6B pencils, thick and porous-tipped pens, notebooks with black and spaced staves, etc.) software with screen magnifiers, programmes with voice synthesis, among others (MIRANDA; PAULA, 2010).

In terms of teaching resources, there is a need to adapt activities that are predominantly visual, as well as teaching methodologies that require the use of this sense. To this end, it is possible to produce teaching materials and games using, above all, scrap metal and low-cost objects:

> It is essential that teaching resources have visual and tactile stimuli that cater for different visual conditions. Therefore, the material must have contrasting colours, textures and appropriate sizes to make it useful and meaningful. The production of teaching resources for blind students should be based [...] on the fidelity of the representation, which should be as accurate as possible in relation to the original model (SEESP/MEC, 2006, p. 15).

Didactic resources that are suitable for working with students with visual impairments are called tactile didactic resources, which must be resistant to exploration by touch, must not cause rejection when handled and, above all, must not offer risks to students. It is important to emphasise that the use of these materials is not restricted to visually impaired students, but also contributes to the teaching of sighted students.

One of the main consequences of visual impairment is the impairment of orientation skills[12] and mobility[13] (OM), so that according to Webster (1976 apud SEESP/MEC, 2006) this limitation "is considered to be the most serious effect of blindness on the individual".

Thus, the act of orientating and moving around visually impaired people is associated with the use of their

[12] Orientation is understood as "the individual's ability to perceive the environment around them, establishing bodily, spatial and temporal relationships with that environment through the remaining senses" (SEESP/MEC, 2006, p. 98).

[13] Mobility is defined as "the individual's ability to move, reacting to internal or external stimuli, in static or dynamic equilibrium" (SEESP/MEC, 2006, p. 99).

remaining senses and the execution of basic mobility techniques, such as protection techniques, information gathering, tracking, alignment, walking sticks and the use of sighted guides (BRUNO, 1993).

Considering what Bruno (1993) points out, the possibility of guiding oneself, getting around, coming and going based on one's own intentionality, contributes to the consolidation of the individual's autonomy. It is necessary for OM skills to be taught to children with VI from the beginning of their development and, therefore, to be part of their educational process. This imposes a responsibility on educational institutions to provide the conditions for making this knowledge a reality:

> [OM teaching] should be started by parents in the home from an early age, followed in formal education by a teacher qualified in the education of the visually impaired. This programme should be completed by a teacher specialising in Orientation and Mobility, when the most advanced techniques are taught, with a view to the most complete and possible independence, effectively favouring the person's real integration into society (SEESP/MEC, 2006, p. 98).

Another important aspect of educating visually impaired students is the need to teach routine activities such as eating, performing and maintaining personal hygiene, looking after objects, etc., since these are learnt mainly through observation and imitation. In this sense, it is important to include Activities of Daily Living (ADLs) in the teaching process for students with visual impairments.

It should be noted that work on ADL begins with the family, and that schools and specialised care centres should complement this teaching. According to Miranda and Paula (2010), ADL provides functional educational opportunities that enable DV students to carry out domestic and self-care tasks independently, which contributes to their well-being and social inclusion.

It is understood that the educational process should not be limited to intellectual knowledge, but should consider other dimensions, such as physical, affective, social, moral and aesthetic (CAVALCANTI, 2002), so that, in fact, citizen education is consolidated, resulting in critical, conscious and active subjects in the social reality of which they are a part.

In this context, teaching visually impaired students has some specificities that require adjustments. However, despite these particularities, it should be emphasised that the main objective of educating students with visual impairments is the same as that proposed for other students. The formation of citizens by educational institutions is independent of the ability to see, which reinforces the need and duty to offer accessible conditions to all.

1.2 CEBRAV: specialised educational care for people with visual impairments

The education of visually impaired students in Goiás has followed the general structuring of Special Education in the state, starting out apart from the regular education system and later being influenced by the integrationist and then inclusive concept. As such, the education of visually impaired students currently takes place in regular schools, which, according to legislation, must provide a series of specific services, spaces and professionals to cater for these individuals.

In addition to regular schools, the education of visually impaired students in Goiás is complemented by specialised educational care, which takes place in specific centres and is characterised by the provision of services aimed at contributing to the educational process and the social inclusion of the visually impaired student. Of particular note is the work carried out at the Brazilian Centre for Rehabilitation and Support for the Visually Impaired (CEBRAV).

In 1998, through a partnership between the Brazilian Association of Educators of the Visually Impaired (ABEDEV), the Ministry of Education (MEC) and the State Department of Education, the Pedagogical Support Centre for the Care of Visually Impaired People (CAP/GO) was inaugurated, which operated the Braille press in the state, with the aim of producing and making accessible didactic-pedagogical material for visually impaired students enrolled in the regular school system.

The context in which CAP/GO was created was characterised by an integrationist educational perspective. In 1999, Goiás' special education policy began to be based on an inclusive concept, which was consolidated, at least in legal terms, in 2000 with the launch of the Education for Diversity in an Inclusive Perspective Programme (Peedi).

CAP remained responsible for braille printing services until 2008, when, as a result of a partnership between the State Department of Education and the Association of the Visually Impaired of the State of Goiás (ADVEG), the Brazilian Centre for Rehabilitation and Support for the Visually Impaired (CEBRAV) was opened. In addition to the activities offered by CAP/GO, it began to provide habilitation, rehabilitation, specialised educational care and ophthalmological assessment and diagnosis services.

The year CEBRAV opened, in line with the broader context of education in Goiás, was marked by the construction/consolidation of specialised educational care centres (CAEE). In 2009, based on the Special Education Department's Operational Guidelines for the Inclusion Support Network (s/d), the characteristics and concepts of this service were defined, as well as the professionals involved and the spaces in which it was carried out.

CEBRAV currently has a multi-professional team and is structured into four service units: the Habilitation and Rehabilitation Unit, the Pedagogical Support Unit, the Human Resources Training Unit and the Graphic Materials Production Unit.

The Habilitation and Rehabilitation Unit, with the aim of promoting the overall development of visually impaired people and contributing to their autonomy and social inclusion, offers the following services: Social Services, Psychology, Functional Vision Assessment, Visual Rehabilitation, Early Intervention, Orientation and Mobility, Motricity, Autonomous and Social Life Activities, Art Therapy, Music Therapy and Ophthalmological Consultation. is the result of an association between ADVEG and the Ophthalmology Reference Centre (CEROF) (GOIÁS, n.d.).

The Pedagogical Support Unit aims to contribute to the educational process of visually impaired students. It is made up of fifteen teachers from different areas and caters for around 160 children and teenagers, in activities that range from Kindergarten to High School. According to the Special Education Management Operational Guidelines for the Inclusion Support Network (n.d.), this unit:

> It provides individual and group ESA for primary school students enrolled in the municipal, state and private education networks. This Unit also carries out the following complementary activities: IT courses, Braille courses for students, dance classes, sports initiation classes, courses for family members in the field of visual impairment, pedagogical assessments, and pedagogical consultancy for teachers in the municipal, state, federal and private schools (GOIÁS, n.d., p. 33-34).

The Human Resources Training Unit is responsible, above all, for improving the professional training of people who deal with visual impairment in their workplaces and practices. As such, it "promotes professional training courses for users, teachers and professionals from related areas" (PACHECO; et. al., 2010, p. 8). The unit's activities include both sighted and visually impaired professionals.

Finally, the Graphic Materials Production Unit, which is responsible for carrying out the activities of the braille press, is responsible for providing enlarged and braille materials in printed and digital format, including the adaptation of books, texts, educational assessments, as well as figures, graphs, tables and captions. This unit provides textbooks in Braille and in enlarged format to visually impaired students in basic education throughout the state of Goiás (SILVA, 2011).

It is interesting to note that the services offered by CEBRAV include children, adolescents and adults with visual impairments and their families, and that the institution is also a training ground for undergraduates from different areas of knowledge. Therefore, given the diversity of activities developed, which also address different aspects of visual impairment, CEBRAV is considered to make an important contribution to the personal, professional and social development of visually impaired people in the state of Goiás.

1.2.1 Supervision at CEBRAV: the role of the teacher

The planning and development of all the activities carried out at CEBRAV were discussed with and had the consent of the supervising teacher[14] , who is one of those responsible for Maths teaching at the institution. The information presented here was obtained from an interview on 17 October 2014.

This teacher's teaching and political work has been very significant since the structuring of education for visually impaired students in Goiás, which has accompanied the political and institutional changes aimed at this service. Her experience with visual impairment, however, began long before her professional career.

"I say that my relationship with visual impairment is the size of my life". He began by explaining his relationship with visual impairment, explaining that since childhood he had lived with a cousin who was blind, through whom he joined the Artisanal Institute for the Blind to teach maths, around 1978/1979.

Later on, she got married and together with her husband (who is blind) became definitively involved in the struggle of visually impaired people in the state of Goiás, "my husband and I, you know that my husband is visually impaired, we have a commitment to the struggle, we can't remain uninvolved, I say that I'm involved from the heart".

In relation to the problem of specific training for teaching students with disabilities, the teacher recounted her experience of being self-taught, not by choice but by necessity, and expressed the importance of professional practice and, consequently, of the students in this process. "My training process was first self-taught and my great teachers were my students. There was no [...] course here, there was nothing. When the first courses arrived, we had already come a long way."

She pointed out that in the current context there are greater opportunities for continuing education, firstly because of the work that has already been consolidated and also because of the partnership between the specialist education centres and the university. Looking specifically at CEBRAV, the teacher pointed out that this partnership, while contributing to the initial training of future teachers, also helps with theoretical discussions and the adoption of methodologies by the institution's professionals.

While criticising the current inclusion policy, he highlighted another important aspect of the partnership between universities and specialised education centres, since the former not only contributes to the initial and continuing training of teachers, but also strengthens the political struggle in defence of inclusive education in the state.

With regard to research and publication by CEBRAV professionals, the teacher said that the process of continuing training has intensified in recent years, with more frequent participation in specialisations, courses and events, as well as the structuring of an internal study group. With regard to publishing, she said it was less common, but that closer ties with the university were a possibility for overcoming this aspect.

Considering the current context, she was asked about the importance of the work carried out at CEBRAV for DV students and their families. The teacher replied, "I feel the following: the children learn at CEBRAV", demonstrating the need for education systems to value the various aspects of education, without shying away from the intellectual

[14] It should be emphasised that contact with this teacher began during my monograph research, and the possibility of working under her supervision again was one of the factors that led me to choose CEBRAV as my field of research.

construction of these subjects.

Finally, when asked about her future expectations for teaching visually impaired students, she said that the difficulties and weaknesses do not suppress her belief and hope for improvement, "it's the history of humanity, we move forward".

CHAPTER 2

THE PLACE WHERE I LIVE FITS IN MY HAND: SCHOOL CARTOGRAPHY, TACTILE CARTOGRAPHY AND THE CONCEPT OF CARTOGRAPHIC SCALE

Geography's scientific identity, as well as its contribution to the development of society, is structured around discussions and understandings of geographical space. Various definitions and perspectives are drawn up on this, which are articulated with other concepts (territory, region, landscape, place), guiding the geographical analysis of reality.

At the same time as it influences socio-environmental dynamics, geographical space is constructed and reconstructed by them. According to Corrêa (2008, p. 28):

> In reality, space organised by man [geographical space] plays a role in society, conditioning it, sharing in the complex process of social existence and reproduction.

Geography as a curricular component shares responsibility for analysing and understanding space with the academic science of geography. Enabling students to understand this concept in its multiple dimensions (cultural, economic, environmental and social) is the main contribution of school geography to the development of citizenship (CASTELLAR, 2005).

The ability to analyse space is closely linked to the construction of spatial reasoning, which is not limited to the locational aspect, but proposes reflection on locations and their influence on social and environmental processes. This reasoning is related to everyday spatialities and the resulting knowledge. According to Cavalcanti (2012, p. 45-46):

> In their daily activities, students and teachers construct geography, [...] they build places, produce space, delimit their territories. Thus, they are forming everyday spatialities [...] It is up to the school to work with this knowledge, discussing, expanding and altering the quality of the students' practices, in the sense of a reflective and critical practice.

In addition to understanding the fundamental concepts of Geography, the construction of spatial reasoning involves understanding cartographic knowledge. Cartography is a "language, a code-system of communication that is essential in all spheres of geography learning" (CASTELLAR, 2005, p. 216). It makes it possible to represent, interpret and understand objects and phenomena through the lens of spatiality, making it possible to build basic skills in spatial analysis.

Considering cartography as an indispensable language for geography education, Passini, Almeida and Martinelli (1999) state that:

> By unravelling cartographic language, the subject can improve their ability to read and understand maps. At the same time, they will be equipping themselves to use this skill to better read and understand geographical space, the world. This process, which is also infinite, goes beyond decoding. At each step, it mainly leads to an improvement in cognitive structures, making it possible to read at different levels of complexity: new relationships, comparisons, orderings, critical analyses, explanations and propositions (p. 126-127).

Cartographic language is based on graphic representation, which is defined as a visual communication language with a monosemic character, i.e. universal and unambiguous (MARTINELLI, 2003a). The main function of this representation is to guide the preparation of cartographic products:

> Transcribe the three fundamental relationships - diversity, order and proportionality - between objects by visual relationships of the same nature. Thus, diversity will be transcribed by a visual diversity; order by a visual order and proportionality by a visual proportionality (MARTINELLI, 2003b, p. 13).

Table 2 shows the graphical transcription of the three fundamental relationships.

Relationships between Objects	Concepts	Graphic Transcription

Notebook	Pencil	Rubber	≠	▲ ● ✚
Gold medal	Silver medal	Bronze medal	O	● ○ ⊙
1kg rice	4kg of rice	16kg of rice	Q	▪ ■ █

Chart 2 - Graphical transcription of the three fundamental relationships
Source: Martinelli (2003a, p. 14)

Information about the fundamental relationships of objects and phenomena is represented through three modes of implantation: point, linear and zonal and expressed by visual variables: size, texture, value, colour, orientation and shape, as can be seen in Chart 3.

Size		Small, medium, large with proportion
Value	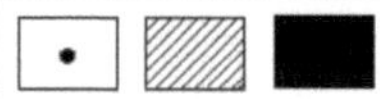	Light, medium, dark
Granulation		Fine, medium, coarse texture
Colour		Red, yellow, green
Guidance		Horizontal, vertical, oblique
Shape		Rectangle, circle, star polygon

Chart 3 - Six sensitive visual modulations Source: Martinelli (2003a, p. 14)

Visual variables have specific perceptual properties, defined as Selective Perception: the eye is able to isolate the elements (visual variable - orientation); Ordered Perception: the categories are ordered spontaneously (visual variable - value: from light to dark); Quantitative Perception: the relationship of proportion is immediate (visual variable - size); Dissociative Perception: moving different sizes away from the eye, they disappear successively (visual variable - size); and Associative Perception: the categories are confused, moving them away from the eye, they don't disappear (visual variable - orientation) (MARTINELLI, 2003a).

Graphic representation, together with reduction and projection, structures and guides the production of maps, which are understood as "a coded representation of a given real space [...] a communication model that uses a complex semiotic system" (ALMEIDA; PASSINI, 1998, p. 15). Thus, it is essential to approach this knowledge in geography teaching, as the representation of space through maps contributes to the development of spatial reasoning and enables students to improve their understanding of space.

The issue of building cartographic knowledge in the context of school geography was addressed by Simielli (1999), who systematised a proposal whose aim is to train critical readers and conscious mappers at the end of the teaching-learning process. This methodology therefore advocates the use of cartographic products that have already been finalised and produced by the students themselves.

Simielli (1999) argues that the beginning of work with cartography at school should make it possible to understand cartographic language, and it is pertinent to consider the study of concrete space, that which is closest to the student, such as the classroom, the school, the neighbourhood, etc.

Subsequently, the work is guided by two complementary axes. The first axis is characterised by the use and exploration of cartographic products that have already been produced and aims to train critical readers who will be able to analyse and interpret different representations (SIMIELLI, 1999). In this axis, the student will analyse representations that have greater cartographic rigour, becoming familiar with symbols, conventions and standards particular to cartography.

The second axis aims to train students as conscious mappers. It presupposes the effective participation of this individual in the processes of drawing up and making cartographic representations. Simielli (1999) argues that three-dimensional and two-dimensional perceptions can be addressed in this axis, and to this end the possibility of working with models, sketches and mental maps is presented.

It is essential to understand that the two axes proposed for teaching cartography are not competing processes, nor are they hierarchical. Simielli (1999) emphasises this:

> Both the first and second axes do not compete with each other, nor does one lead to better results than the other. Both axes are correct, both axes produce excellent results. What we have are teachers who are better suited to working with the first axis and teachers who are better suited to working with the second axis and, depending on the subject, the teacher may work better with one axis and, in other subjects, opt for the other axis (p. 107).

The proposal to train students to be critical readers and conscious mappers through the teaching of cartographic knowledge presupposes that schools articulate educational practices based on Cartesian and non-Cartesian approaches to cartography. These, like the axes presented by Simielli (1999), "represent complementary (not opposing!) teaching strategies" (SEEMANN, 2011, p. 40).

In relation to Cartesian cartography, in which representations are constructed based on conventions and standardisations, in addition to the relevance of maps, charts and plans, the importance of working with municipal school atlases stands out. These are defined as "collections of maps, at different scales, which represent various aspects of a given space" (BUENO, 2008, p. 20), and when they acquire a pedagogical focus, they become didactic materials in the teaching-learning process and are called school atlases.

The use of school atlases in geography teaching is considered to favour the understanding of fundamental concepts and the development of skills such as observation, recording, representation, analysis, interpretation and communication. Furthermore, when this material is made up of data from the municipality, representing local objects and phenomena, its contribution to geography teaching is even more significant.

Bueno (2008) states that "the primary goal of municipal school atlases is to enable the study of place as an everyday space" (p. 28), going beyond its locational perspective. This approach re-dimensions both the process of analysing, interpreting and understanding cartographic data and the meaning that people give to place itself. Therefore:

> When working with atlases, the student's and teacher's gaze is directed towards the printed conventions and symbols, but also towards the indices of natural and cultural phenomena that occur in the municipality, in a real geographical context. The dialogue between the subjects involved in the process creates a link between representation and meaning (SILVA, n.d., p. 3).

Associated with the characteristics, definitions and contributions of Cartesian Cartography are the perspectives of non-Cartesian Cartographies, which becomes absolutely coherent when we recognise how complex the practices of spatial perception and representation are.

We highlight the contribution of mental maps, which according to Seemann (2013) are "a graphic autobiography, a spatial memorial of a person" (p. 106). In this way, such representations are based on particular experiences of everyday socio-spatial practices, making the choice of what to represent and how to represent pass through the sieve of subjectivity, which invariably distances itself from cartographic standards.

In addition, mental maps, which represent aspects that are usually neglected in official maps, bring new possibilities to cartography. These include valuing the process of drawing up and producing representations and not just the final product, and giving new meaning to reading and interpreting products that have already been drawn up. Seemann (2013) explains:

> Personal experience and associations with places give a different meaning to the map with its austere appearance, its precise scale and its cartographic conventions. The map ceases to be a mere document for localisation and orientation or a utilitarian product for the authorities (town hall, government). The personal reading of the map does not obey the rules of geometry, precision and technical production: it is done through the private and intimate code of memory (p. 106).

In this sense, the process of map reading requires an understanding of cartographic culture, which proposes the investigation of the socio-cultural contexts that underpin the conceptions, definitions and characteristics used in the various cartographic representations.

Considering the influence of school cartography on the implementation of spatial reasoning, it is argued that the interrelationship between the various cartographic concepts favours the construction of such reasoning. We will therefore discuss a very significant element from the perspective of Cartesian cartography, which also influences, albeit in a different way, the reflections of non-Cartesian cartographies: the concept of cartographic scale.

2.1 The cartographic scale and its place in school cartography

The concept of cartographic scale assumes great importance in the construction of cartographic knowledge, influencing both the analysis and interpretation of representations, as well as the process of preparing and making them. Cartographic scale thus contributes to the critical reading and understanding of geographical space, influencing people's daily lives:

> Mastering the concept of scale makes everyday life much easier because it helps us to perceive distances and therefore the time likely to be spent travelling and allows us to situate ourselves in space at various levels, local, regional and even worldwide (LE SANN, 1984, p. 56).

The cartographic scale is defined by a ratio established between the representation and the space/object represented, i.e. between the "distance on the map (generally 1 cm) and the equivalent distance on the ground (x cm)" (LE SANN, 1984, p. 56), constituting a concept that requires a high level of abstraction.

Although the cartographic scale is essentially abstract, relating to numerical quantification, its constitution is outlined by the perception of the concrete and non-quantifiable, and in both structures (non-quantifiable and quantifiable) it is based on the notion of proportionality .[15]

In addition to the relationship of proportion, Le Sann (2007) argues that the construction of the concept of cartographic scale is linked to two other stages: representation and comparison. The representation stage corresponds to the perception and signification of reality without necessarily establishing measurements. Comparison, on the other hand, is associated with variation in size (reduction and enlargement), Le Sann (2007) states that:

> The notion of scale applies to the size of the representation: it can be transposed, the size doesn't change, or it can be transformed, increased or decreased depending on the scale. The concept of size is based on the perception of differences by sight or touch and is developed at this stage of comparison. In order to accurately compare

[15] The construction of proportional reasoning will be covered in topic 2.2 of this chapter.

variations in size, the concepts of length and width must be correctly assimilated (p. 16).

Proportion, on the other hand, refers at first to non-quantification (larger/smaller/equal) and then to quantification (how many times larger or smaller) (LE SANN, 2007). It is therefore important not to limit the concept of cartographic scale to the exactitude of mathematical calculations, but rather to consider the relationships of proportionality that exist in everyday situations and that can be perceived by analysing and reflecting on the space we live in, the elements that make it up and their relationships.

Regarding proportionality in everyday life, Pissinati and Archela (2007) state that:

> In their daily lives, human beings don't even realise how much they use proportionality. Photographs, the cinema or TV screen, toys (such as cars and dolls) or artistic landscape paintings connect us with real-sized elements, but we don't even think about it. In fact, when it comes to scale, the correlation between it and proportionality isn't even immediate in our minds. We also use proportionality when it comes to inserting one object into another, seeing in advance that this will be possible (p. 182).

Encouraging students to perceive and represent objects of varying sizes, using observation, memory and comparison between the real and the image, helps to improve their notion of proportionality, broaden their operative skills of comparison, quantification and classification and, consequently, their construction of the concept of cartographic scale (CASTELLAR, 2005).

As well as understanding the non-quantifiable and quantifiable dimensions of the concept of cartographic scale, it is necessary to address the technical specificities of this concept in the context of cartographic products. Thus, scale can be represented in two different ways, "by a fraction (e.g. 1/10,000), in which case it is called a 'numerical scale', or by a drawing [...] which is the 'graphic scale'" (LE SANN, 1984, p. 57). It is important to emphasise that the graphic scale remains accurate when cartographic products are enlarged or reduced.

In numerical form, the "1" refers to the dimensions of the map and the number after the division symbol corresponds to the actual dimensions. Depending on the relationship between the numerator and denominator, the scale can be defined as natural, enlarged or reduced:

> When the Numerator and Denominator are equal, the scale is said to be NATURAL. The Title of a natural scale will be "1:1" [...] When the Denominator is smaller than the , the scale is said to be ENLARGED. The Title of an enlarged scale could be, for example, "1: 0.5" [...] When the Denominator is greater than the Numerator, we have a REDUCED scale. An example of the Title of a scale of this type could be "1: 50 000" (DUARTE, 1983, p. 34).

The size of the scale influences the type of representation and its characteristics. Table 4 shows the specific characteristics of cartographic products according to the scale defined.

Cartographic product	Scale	Features
Floor plans	1:100 (large scale)	It portrays the details well
City plans or cadastral plans	From 1:500 to 1:5,000 (large scale)	Details the essential lines without deformation, showing the exact location, linear and areal dimensions of buildings
Topographical maps	From 1:20,000 to 1: 250,000 (large scale)	This representation is limited to geographical coordinates and rarely to political boundaries
Chorographs (smaller scale charts)	1:500,000 (small scale)	Provides an overview of a region
Maps	1:1,000,000 (small scale)	Fewer details and the limits of the represented area are the political limits

Planisphere (general charts)	1:205,000,000 (small scale)	They represent the whole world

Table 4 - Classification of cartographic products according to scale Source: Based on Almeida and Passini (1998)

Recognising the importance of the concept of cartographic scale to the teaching of cartography and geography and considering that the notion of proportionality is fundamental to its structuring, it is necessary to discuss it in more detail.

2.2 The notion of proportionality and spatial relations

The cartographic scale is a ratio of proportion between the representation and the real thing, with distance (size) being the variable in which the proportionality relationship is established. Although the notion of proportion is not only linked to the concept of scale, considering it in this context requires understanding the structuring of proportional reasoning through a mathematical lens, associating it with spatial relationships.

According to Spinillo (1994), understanding proportionality is generally reduced to numerical quantification (calculations), the use of the rule of three algorithm and/or symbolic representation (y/a = x/b), which according to the author, "does not guarantee an understanding of the meaning of the relationships involved in the concept" (SPINILLO, 1994, p. 110).

Thus, the formation of proportional reasoning involves the ability to establish relationships between values or quantities (discrete or continuous[16]), but its construction is based on proto-quantitative schemes:

> They allow children to reason and establish relationships without the need to quantify them numerically. Later in development, this knowledge is transformed into quantified and mathematically exact forms of representation. Proto-quantitative judgements, although not numerical, can include the essential principles of a concept; in the case of proportion, these principles refer to the relationships that need to be established either numerically or non-numerically (SPINILLO, 2002).

The structuring relationships of proportional reasoning are basically of two types: first-order relationships, which are subdivided into part/part and part/whole relationships; and second-order relationships. Therefore, proportion judgements begin with part/part relationships, which subsidise the establishment of part/whole relationships and second-order relationships, and it is interesting to note that this process is interrelated and non-linear.

First-order relationships correspond to proportional judgements made about the variables in a set. First the parts are compared (part/part) and then the part with the whole of that set (part/whole), at which stage each variable (length, width, distance, etc.) is generally considered in a particular way. Second-order relationships consist of comparing first-order relationships with each other (relationship of relationships) (RUIZ; CARVALHO, 1980).

In the context of non-numerical forms, proportional reasoning relationships are established based on estimates, perceptual skills and qualitative comparisons such as: 'more/greater than', 'less/less than' and 'equal to'. In relation to semi-numerical and numerical forms, these relationships are established through simplified and later complex quantifications (BRYANT, 1974 apud SPIMILLO, 2002).

Spinillo (1994) describes an activity that exemplifies first-order relations, of the part/part and part/whole types, and second-order relations:

> Let's take two sets of marbles as an example, one consisting of three blue marbles and five yellow marbles and the other of three blue marbles and three yellow marbles. To decide which of the two sets has the highest proportion of blue marbles, the child first needs to compare the number of blue marbles in each set in relation

[16] It should be noted that discrete variables are those that correspond to integer values, for example, number of children, and continuous variables relate to fractional values, for example, weight, height, etc (AGUIAR, 1993).

> to the number of yellow marbles (3 blue marbles to 5 yellow marbles, and 3 blue marbles to 3 yellow marbles) or compare the number of blue marbles to the number of marbles in each set (3/8 of the marbles are blue and 3/6 of the marbles are blue). These are first-order relationships. [...] After establishing the first-order relationships, the child then needs to compare them with each other to decide which of the two sets has a higher proportion of blue marbles. This comparison between the first-order relations is the second-order relation (relation between relations) (p. 110).

Linked to first and second order relationships, proportional reasoning is structured by recognising the equivalence between different situations, which is made possible by thinking in relative terms rather than in absolute terms. In this way, the variables in a set (quantity, distance, height, width, etc.) must be analysed in the context of proportionality.

Some authors argue that the difficulties in building proportional reasoning are mainly concentrated in establishing second-order relationships and relative judgements (RUIZ; CARVALHO, 1980); (SPINILLO, 1994; 2002). To this end, it is recommended that the construction of proportional reasoning be based on the concrete, as Ruiz and Carvalho (1980) exemplify "it is easier for a child to understand the relationship between the length of a stick and the size of its shadow than it is to understand simple numerical relationships".

Furthermore, it is considered important to provide teaching-learning situations in which students can apply the concept of proportionality and judge the equivalence or non-equivalence of different variables and sets. This proposal reinforces the importance of bringing proportional reasoning closer to the student's reality and avoiding early algorithmisation.

Specifically considering the cartographic scale, in which the relationship of proportion affects the variable distance (size), which in turn relates to space and its representation, it is necessary to reflect on how spatial notions and relationships are constructed.

According to Piaget and Inhelder (1993), spatial relationships develop on two levels: the perceptual or sensorimotor level and the representational or intellectual level, so that:

> The perceptual or sensory-motor structures are the starting point, since they are formed from direct contact with objects and it is later that the representative construction of space takes place, since it evokes the image of the object in its absence. It only occurs long after the appearance of language and representation, which occur more or less simultaneously with perception and motor skills (MONTOITO; LEIVAS, 2012, p. 27).

In this way, perceptible space is constructed from the beginning of existence, linked to the progress of perception[17] , action and motricity. Starting from an egocentric perception, perceptible space is characterised, firstly, by the lack of coordination of sensory spaces (taste, visual, auditory, postural), the lack of differentiation between the movements of objects and those of the subject and the lack of evaluation of distances and relative positions.

As egocentrism gives way to decentration, characterised by the coordination of different points of view, perceptual activity (associating perception with movement) is improved, enabling comparisons and transpositions of perspectives. And so the representative space is formed.

It should be emphasised that the construction of visually controlled manipulation schemes is of great importance in the process of transforming perceptual space, since it leads to the analysis of figures and shapes from a spatial point of view. Thus, the absence of the sense of sight interferes significantly, because as Piaget and Inhelder (1993) state:

> Exploration by eye is easier than by hand, because a visual centring encompasses many more simultaneous elements than a tactile centring. Thus, visual forms are constructed more quickly than tactile forms (p. 40).

[17] For Piaget and Inhelder (1993), perception is a "certain system of relations, organised into an immediate totality, but the balance of this totality depends not only on real relations, but also on virtual relations, in other words, they refer to previous or possible perceptions. Thus, every perception implies a sensory-motor system, which applies the result of all previous constructions to the current situation" (p. 30).

In the context of visual absence, Piaget and Inhelder (1993) emphasise the importance of stereognostic perception, which corresponds to translating the tactile-kinesthetic perception of a certain invisible object into a visual spatial image. According to these authors, this perception makes it possible to analyse both the construction of figurative intuitions and the mechanisms of tactile perception.

Representational space, on the other hand, comes after the emergence of language and coincides with the beginning of imagery and intuitive thinking. According to Piaget and Inhelder (1993) "Representation is obliged to reconstruct space from the most elementary intuitions, [...] applying them, in part, to projective and metric figures higher than the level of these primitive relations" (p. 18).

Thus, representative space corresponds to "an internalised action and not simply the imagination of any external data, the result of an action" (PIAGET; INHELDER, 1993 p. 474). This space is made up of two elements: the intuitive space, resulting from the internalisation of spatial actions carried out at a perceptual-motor level, which at this point are static and irreversible; and the operative space, which is characterised by the operationalisation of the intuitive space, with the representations being mobile and reversible (PAGANELLI; ANTUNES; SOIHET, 1981).

Piaget (1979) emphasises that the representative space makes it possible to carry out reversible operations, coordinate perspectives, maintain objects and establish relationships between them and no longer as a function of the subject; this capacity is called Secondary Circular Reaction. This space is therefore characterised by the simultaneous construction of projective figures (coordination of points of view on objects) and Euclidean figures (by the constancy of the dimensions attributed to the object).

The construction of perceptual and representational spaces is linked to the development of spatial relations, which are of three types, topological, projective and Euclidean, with the latter two "appearing later than the most elementary spatial relations [so that] spatial perception involves a progressive construction" (PIAGET; INHELDER, 1993, p. 20).

Topological relationships concern the properties of objects in themselves. According to Paganelli, Antunes and Soihet (1981) these relations "do not consider rigid shapes, distances, lines and angles and are the most elementary for the construction and representation of space" (p. 2). In this way, elementary spatial relationships are characterised as:

- Neighbourhood: corresponds to the proximity of perceived elements in the same field;
- Separation: distinguishing between two neighbouring elements;
- Order (or succession): is established between elements that are both neighbouring and separate, and is constituted from the sequence of elements and movements;
- Circumscription (envelopment): is established in the different dimensions (one dimension - between; two dimensions - framed; three dimensions - interiority);
- Continuity: is perceived through displacement, being modified by the improvement of relations of neighbourhood and separation (PIAGET; INHELDER, 1993).

Projective space begins, psychologically, when the object or its figure are no longer considered simply in themselves or as a function of the subject, but relative to different points of view, in other words, these relationships go through the coordination, firstly sensory-motor, and then operative, of perspectives on the perception and representation of objects.

The change of perspective, according to Piaget and Inhelder (1993), corresponds to the various forms that an object takes on as it changes position in space. There is a correspondence between the perceptual point of view and the others, reflected in the transformation of the relationships of left vs. right, front vs. back and up vs. down:

> An object considered from a certain point of view: there are then other elements to its left or right and "between"

> which it is placed from the subject's point of view. [...] there are, on the other hand, elements situated above and below it, from the subject's point of view, and these second relationships characterise a second dimension according to height; finally, there are elements situated in front of or behind it along the line linking it to the subject's point of view, and these new relationships characterise a third dimension according to depth (p. 206).

Finally, Euclidean or Metric Relations "are based on the notion of distance and allow objects to be placed in relation to each other, considering a fixed reference system" (PAGANELLI; ANTUNES; SOIHET, 1981, p. 5-6). In this way, the development of these relationships allows for the conservation of distances and lengths (in perceptual and representative terms), in one, two and three dimensions.

It is interesting to note that Euclidean relationships are established from the informal construction of the concept of measurement, in other words, subjects understand the notion of measurement before developing awareness of dimension and quantification. Thus, at a very young age, children "already know how to compare their height with that of their peers, are aware that they can't reach an object that is much higher than them, etc" (MONTOITO; LEIVAS, 2012, p. 30).

The construction of proportional reasoning was discussed by Piaget, above all, in relation to geometric proportions, which are linked to the notions of angle and similarity of form, resulting from Euclidean spatial relations. However, this author argues that the development of proportionality is influenced and permeates topological and projective relations, in other words, "the proportionality of dimensional relations is engendered [...] by the similarity previously established thanks to qualitative relations" (PIAGET; INHELDER, 1993, p. 359).

In addition, the construction of proportionality is linked to the comparison procedure, which is established, in principle, through the perception of objects and figures, and is not linked to dimensional notions. Therefore, in topological relationships, it is already possible to establish comparisons between objects through perceptual judgements (PIAGET; INHELDER, 1993).

After the purely perceptual judgements, the awareness of parallelism, equality of angles and simple dimensional relationships begins, which vary according to the perspective of observation. Subsequently, dimensional relationships about proportion are applied to one dimension (usually length) and finally to the other dimensions of the figure.

Like perception, the representation of proportional reasoning is based on perceptual judgement, through which only shape is considered. This is followed by an awareness of parallelism and the equality of angles, and finally the dimensional perspective, which enables the construction of representative proportionality (PIAGET; INHELDER, 1993).

Linking the relationship of proportionality to the concept of cartographic scale, which in turn is part of the context of school geography, requires that reflection on space is not limited to the physical aspect, but that it considers other spatial dimensions. In this sense, we propose a discussion on the geographical concept of place.

The construction of the concept of cartographic scale is related to the perception and relationship of quantifiable proportion established between space and its representation, and is therefore linked to the physical and dimensional scope of space. It is argued, however, that the inclusion of the concept of cartographic scale in the context of geography teaching requires consideration of other spatial dimensions, since geographical space is not only constituted by materiality. The aim is therefore to discuss the concept of place, understood as the space of everyday life and relationships.

The discussion of place based on these dimensions has been carried out, above all, by phenomenologically-based Humanist Geography[18] , which has sought to overcome the locational conception associated with this concept,

[18] In geographical science, "the phenomenological method would be used to make a rigorous description of the lived world of human experience and thus, through intentionality, recognise the '**essences**' of the perceptual

understanding that when space is experienced and gains meaning for the individuals who experience it, then it becomes a place (BUTTIMER, 1976).

Place is the space that is appropriated, experienced and felt, in which daily actions and common facts of life take place. Tuan (1983, p. 158) argues that "simple events can over time develop into a deep feeling for place".

The process of spatial signification and identification is influenced by the intersubjective relationships established with other subjects and with the place itself, which occur "at the moment when the body [...] makes contact with the exterior and localises the other, communicating with other men and getting to know other situations" (HOLZER, 1997, p. 79).

In addition to inter-subjective relationships, the sense of place is built up by the experiences[19] that people have and which, in turn, are linked to the passage of time. Thus, awareness of place is influenced by the temporal dimensions of past and present, because elements from previous generations are inscribed in places, enabling those who experience it in the present to understand it through memories of previous times.

Lowenthal (1995, p. 83) argues that "remembering the past is crucial to our sense of identity: knowing what we were confirms what we are", and that this process occurs both individually and collectively and contributes to the formation of both individual and collective identities. The present is also indispensable to a sense of place, because as Tuan (1983, p. 218) states "it is our experiential reality, the sensitive point of existence".

Distance is another factor that influences the constitution of place, and is not limited to the aspect of objective measurement, but is linked to the experiences of the subjects. Dardel (2011, p. 10) proposes that "distance is experienced not as a quantity, but as a quality expressed in terms of near or far. What is close is what can be approached without effort, what is far away requires effort and, implicitly, a desire to get closer".

Tuan (1980) calls the sentimental relationship that individuals have with the places where they live Topophilia, which is influenced at the same time by individual personality and cultural aspects, since:

> Sensory stimuli are potentially infinite: what we decide to pay attention to [...] is an accident of individual temperament, purpose and the cultural forces at work at a given time (p. 129).

Therefore, place is where everyday relationships and actions take place, as it allows the body to experience space, appropriate it, feel it and become attached to it. To the extent that the individual confers meanings on the place, derived from subjectivity and experience, the more the production of place is linked to the production of life, and the opposite is also true.

2.3 Tactile cartography and geography teaching: theoretical and methodological foundations

Historically, cartographic products were built with the sense of sight in mind, which made it impossible for visually impaired people to access the information represented. Tactile Cartography was then structured as a branch of Cartography, whose aim is to help make cartographic knowledge accessible to the perception of visually impaired people, and is defined as:

> The science, art and technique of transposing visual information in such a way that the result is a document that can be used by visually impaired people (CARMO, 2009, p. 46-47).

---

structure" (HOLZER, 1993, p. 140) (emphasis added).

[19] Experience is understood as the way in which someone knows and constructs reality (HOLZER, 1997).

In Brazil, academic research into tactile cartography began in the Department of Geography at the University of São Paulo, with Professor Regina Araujo de Almeida Vasconcellos' doctoral thesis[20] , defended in 1993. The aim of this research was to develop and evaluate a tactile (and visual) graphic language suitable for visually impaired people.

Currently, professors at different higher education institutions are carrying out research, supervising work and participating in groups dealing with the subject of tactile cartography. Of particular note are the studies carried out at the Teaching and Didactic Material Laboratory (LEMAD) at the University of São Paulo and the Tactile and School Cartography Laboratory (LABTATE) at the Federal University of Santa Catarina.

While in universities Tactile Cartography is established as a potential field for research, in schools it is seen as a possibility to make the processes of representation, analysis and interpretation of spatial information accessible to visually impaired students. Regarding the importance of Tactile Cartography, Vasconcellos (1993) states:

> In the case of visually impaired students, the importance of maps is even greater. Diagrams, illustrations, models and maps, despite being abstractions of reality, manage to concretise space, synthesising the information to be perceived by touch. Maps can be used for localisation, orientation and locomotion, along with a compass, at the scale of the building. For visually impaired people, these resources can be used to help them get around in everyday life, at school or in the neighbourhood. In this way, the map is fundamental to the user's perception and construction of space, mainly because they cannot capture spatial information through sight (p. 31).

In addition to understanding geographic and cartographic knowledge, working with Tactile Cartography in Geography classes is justified by the presence, although still not very significant, of tactile graphic representations outside the school environment. Sena (2008) argues that these representations help the orientation and mobility processes of visually impaired people, and that it is essential to equip them to interpret these resources.

It is also important to emphasise that the use of tactile cartographic representations is not limited to DV students, because when associated with visual language, it is also suitable for sighted students. The use of materials that are meaningful to all students is considered to help consolidate more inclusive teaching-learning situations.

Given this context, in which Tactile Cartography is relevant in various spheres, it is interesting to reflect on the specificities of this cartographic process, which, according to Almeida (2011), coincides with conventional cartography:

> It starts with reality (geographical space) and goes through several stages: transformation (from three to two-dimensional, from spherical to flat surface through projections), reduction (scale) and generalisation, codification (graphic and cartographic language), construction and reproduction (p. 147).

According to this author, the most important stage in the whole process is the graphic solution of tactile representations (ALMEIDA, 2011), since graphic language is essentially visual. Thus, based on Graphic Semiology, Vasconcellos (1993) proposed adapting visual graphic language to tactile visual graphic language (Figure 2).

[20] After her doctoral thesis, Professor Regina Araujo de Almeida Vasconcellos changed her name to Regina Araujo de Almeida.

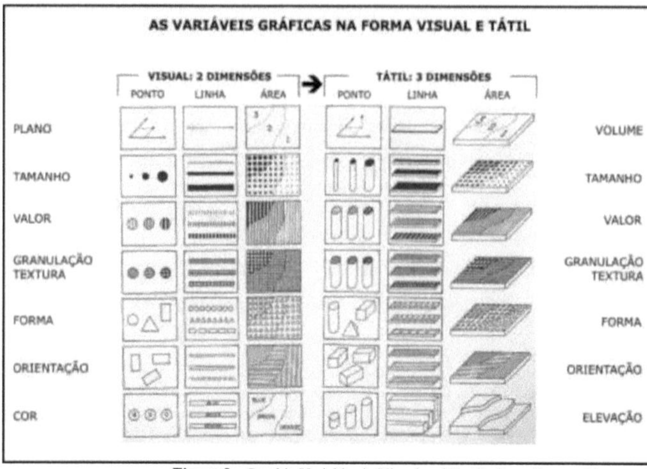

Figure 2 - Graphic Variables in Visual and Tactile Form Source: Vasconcellos (1993)

Bertin (1967) established that graphic language, understood as a system of signs, is made up of meaning (concept) and signifier (graphic image). In this way, the relationships of similarity/diversity, order and proportionality (meanings) are represented by the visual variables (signifiers), which vary according to the form of implantation, which can be punctual, linear or zonal.

The suitability of visual variables for tactile perception requires them to be adapted for use in the third dimension, i.e. in relief. Colour is the only visual variable that cannot be used in tactile form and is therefore replaced by the elevation variable (difference in heights). Therefore, the possible tactile variables are: size, value, shape, grain/texture, orientation and elevation (VASCONCELLOS, 1993).

It should be noted that colour is not discarded in Tactile Cartography. In order to facilitate the use of the visual residue by users with low vision, strong and contrasting colours are used. With regard to this discussion, Carmo (2009) states that:

> Colour is a visual variable, which is why it is used as a complement for people with residual vision [...]. It is very important in the case of visually impaired people who can perceive shades and contrasts, using strong, complementary colours to differentiate symbols, making it easier to use the tactile map (p. 56).

As with conventional cartography, Tactile Cartography representations depend on the nature of the information, which can be qualitative, quantitative or ordered. This influences the selection of the correct variables and, consequently, the quality of the representation constructed.

With regard to the forms of implantation (point, linear and zonal), it is essential to consider the specificities of tactile perception, which is a sequential process, i.e. the apprehension of information about objects, environments and the cartographic representation itself is carried out in a fragmented way.

It is therefore essential to limit the number of point and linear symbols, giving preference to those with simplified shapes that contrast with each other and their surroundings. Carmo (2009) also warns that "the proximity between signs can also make it difficult to read tactile graphic representations" (p. 62), as it leads the visually impaired user to interpret them as corresponding to a single piece of information.

When analysing point symbols in particular, they shouldn't be too big, so that you have to run your finger around them to explore them, and the opposite should also be avoided, i.e. point symbols that are too small.

Ventorini (2007) explains that:

> Sight can easily distinguish the differences between small geometric shapes (cartographic symbols, such as those used to represent minerals), but touch cannot. Elements represented with small dimensions can have their shapes confused by the blind user, because they can't trace their contours with their finger, but only feel them in the form of dots (circles) (p. 71).

With regard to area representations, there are various materials that can be used and the choice of textures and different combinations of these make for more appropriate tactile cartographic representations. The caveat is that the materials selected should not be hazardous or repulsive to users (CARMO, 2009); (SENA, 2008); (VASCONCELLOS, 1993).

Given the need to consider the number of symbols in tactile cartographic representations, it is essential to reflect on the amount of information represented. Vasconcellos (1993) states that "it is important [...] never to overload the map" (p. 47).

In this way, only the main information is represented, as supplementary information is considered to be overload that hinders the reading and interpretation process for visually impaired users. Sena (2008) gives an example:

> A climate map of Brazil made with different textures will not have the political-administrative division, as this information superimposed on the climate information could confuse the visually impaired user when delimiting the areas corresponding to each climate type. On a printed map, the same division can be represented with a softer line or even in the background (p. 44).

The development of various cartographic products makes it possible to solve the problem of the concentration of information and symbols in tactile representations. Vasconcellos (1993) proposes that map collections are a suitable alternative for cartography work with visually impaired users.

It should be noted that understanding the representations that make up a collection is enhanced when the symbols are built using the same textures and colours. Maintaining the representation of information that recurs in various cartographic products favours the correlation between signifier and signified, thus facilitating understanding of the sign.

To the extent that the resolution of touch is lower than that of sight, geometric precision and cartographic rigour are relativised in Tactile Cartography, which results in representations with a higher degree of generalisations, omissions, exaggerations and distortions. Thus, "what is right and desirable for a conventional map is not effective for representation in tactile format" (ALMEIDA, 2011, p. 156).

The size of tactile cartographic representations also influences their reading and interpretation processes. Ventorini (2007) discussed the specificities of this aspect, based on research carried out by Rowell and Ungar. According to these authors, size varies in three respects:

> a) Sizes that make reading easier: the sizes used largely respect the distances considered appropriate between the two hands for exploring the representations, using reference points that make it possible to code and relate the locations of objects on the map;
> b) Suitable sizes for distribution and transport: tactile maps are often damaged when folded or rolled up. In this sense, small maps can be distributed and transported in folders and bags, reducing the risk of damage during transport;
> c) Production medium: the production medium also influences the choice of tactile map size. Most of the machines used to produce high relief copies using microcapsulated paper and plastic for thermoform machines do not use paper larger than A3 (ROWELL; UNGAR, 2003a; 2003b apud VENTORINI, 2007, p. 69-70).

With the aim of facilitating reading and considering the tactile field of exploration, Vasconcellos (1993) recommends that maps, models and tactile graphics should not exceed 50 cm in length and width, with this measure being relativised according to what is intended to be represented, the type of representation and its objectives.

The realisation that tactile cartographic representations are generally explored by hand means that resistance and durability are influential variables in the processes of designing and producing tactile cartography. In this context, it is important to select materials and build representations that are not easily deformed or dismantled.

In addition, the use of low-cost materials is also a concern, in order to make it feasible to produce and use tactile cartographic representations in everyday practices, including schools. Sena (2008) points out that:

> Considering that most schools catering for visually impaired students have few resources to buy materials, the alternative of using various types of paper, reused fabrics and scrap metal in general has expanded the possibilities of improving the quality of the representations used in the classroom (p. 121).

Based on the fundamentals and specificities of tactile cartography, it is possible to construct various representations, including tactile maps and tactile models. Tactile maps are representations made up of tactile language, raised signs, strong contrasting colours and information written in Braille and ink, in enlarged letters. Considering the particularities of DV users, Almeida (2011) differentiates between two types of tactile maps: orientation and mobility maps and geographical maps:

> They are used by people who are blind or have low vision and need support to get around. An orientation map gives an overview of an area, while mobility maps provide detailed information on how to get from one place to another safely. [...] They are large-scale representations of buildings, routes, streets, public transport networks, neighbourhoods and leisure areas (ALMEIDA, 2011, p. 143-144).

As for geographical maps:

> They represent larger areas at smaller scales than orientation and mobility maps. They can be categorised as: general reference tactile maps, for example, political and physical maps [...]. As for topographical maps, the relief needs to be perceived with the aid of models and scale models, but with adaptations (all symbols in tactile form and text in Braille). Thematic maps represent qualitative and quantitative information using tactile graphic language, with techniques also adapted to this format and in a wide variety of formats and scales (ALMEIDA, 2011, p. 144).

Tactile models, on the other hand, are three-dimensional miniatures of objects, buildings, relief shapes, etc. They are made up of textures, raised signs, strong and contrasting colours, information written in Braille and ink, enlarged letters and other miniature tactile elements. Some tactile models have sound devices, which make it possible to combine tactile perception with auditory information (LABTATE website, 2014).

The types of tactile models are defined based on the scale of representation. There are smaller scale models that represent large spaces, such as landforms and geomorphological structures. And models built on large scales, which generally represent spaces close to the subjects, such as the classroom, school, etc.

Defining the type of representation, the selection and organisation of tactile signs and the scale used depends on the nature of the information represented, the objective of the work with the cartographic product and, above all, the characteristics of the user for whom the material is intended.

2.3.1 Methodological steps for designing and making tactile cartographic representations

Although the use of tactile cartographic representations is not restricted to visually impaired people, the process of designing and making these products must prioritise the specificities of tactile perception. As such, tactile cartography has concepts, rules and techniques that are different from those used in conventional cartography.

According to Sena (2008), the process of creating a tactile representation begins with "reflection on the type of information to be represented, the materials available and the level of knowledge and preparation of the student for whom the material is intended" (p. 120).

The way in which visually impaired people perceive, read, analyse and interpret a tactile cartographic representation is influenced by characteristics such as: age, type and time of visual impairment, tactile training, psychological aspects and schooling. It is therefore essential to know the individual for whom the cartographic product is being built.

It should be noted that in the school environment, teachers have the opportunity to understand and systematise the characteristics of their visually impaired students. Given their daily interaction, this understanding can be improved, contributing to the construction of cartographic representations that are increasingly meaningful and suited to the specific characteristics of these individuals.

Before building the representations and in order to understand the tactile perception of the DV student, it is essential to test the materials that will be used. This stage makes it possible to recognise the textures, sizes and shapes that are most easily perceived, differentiated and identified, and in relation to students with low vision, the visualisation of colours and shapes is also assessed.

To make it possible to test the materials, Vasconcellos (1993) proposed, in his doctoral thesis, the creation of a materials notebook, which is made up of various textures and their plastic copies, which subsidise the definition of tactile graphic signs.

After testing the materials, the stage of making tactile cartographic products begins. These are built using specific techniques, which vary according to the type, objectives, materials available and users of the representation. With regard to two-dimensional representations, there are five techniques that have been systematised by Sena (2008).

- Aluminium Representation

Starting with a base map, the representation is drawn in relief on the aluminium foil. This can be done using a specific tool kit or common tools such as an inkless pen, craft tools, sandpaper, etc. According to Sena (2008, p. 97) "this material is ideal for simple drawings, linear representations with little variation in texture [...] and is a resistant matrix for plastic copies". The limitations of this technique are: little elevation of the relief and limited textures (Appendix 1).

- Collage representation

It is characterised as an artisanal technique that is easy to execute and allows the use of various materials, which makes it possible to represent diverse information in the three forms of implantation (punctual, linear and zonal). What's more, the representations resulting from this technique are generally inexpensive, which makes them feasible at .

The final product can be used directly with users or as a matrix for reproduction in plastic, but when used for this purpose, it is necessary to choose heat-resistant materials[21] . Sena (2008) points out that the disadvantages of this technique are, above all, the higher level of generalisations of the representations and the rapid deterioration of the final product (Appendix 2).

- Cold Porcelain Representation

The material used in this technique (biscuit dough) is inexpensive and easily available. It allows a wide variety of representations to be constructed, as it is available in different colours and adheres easily to other materials. Sena (2008) warns that when using plastic "it is important to finish with a layer of white glue which increases resistance and even makes the representation more pleasant to the touch" (p. 101). When used directly with the user, the representation is very durable and cannot be reproduced in plastic copies (Appendix 3).

- Screen Printing Representation

Making cartographic representations using the screen printing technique allows for reproduction in large

[21] According to Sena (2008, p. 100) "some types of plastic (braillon, brailex, PVC or PET) can be used to produce copies of the matrices made using aluminium or gluing techniques. The maps are placed in a Thermophorm machine. The machine uses heat and vacuum to copy the information from the matrix onto the plastic."

quantities, considerably reducing the value of the final product. With regard to limitations, Sena (2008) points out that "they are concentrated in the resolution of the representation, linear implementations cannot have thin thicknesses as they do not allow the ink to pass through the screen; elevations are limited (maximum 2 mm)" (p. 103). In addition, the production process is considered to take longer as it is a craft technique (Appendix 4).

- Representation on Microcapsulated Paper

This technique allows representations to be printed on any copier or inkjet printer, so that "after the ink dries, the paper goes through some heat-producing equipment, which results in the black areas being raised" (SENA, 2008, p. 104). The advantage of this technique is that the cartographic base (or any other image) used in the representation can be built/edited on the computer and the resulting products are resistant and lightweight, enhancing the processes of transporting and using the material.

The variation in textures and heights obtained using this technique is limited, and in addition, only black prints are raised by heat, which intensifies the need to generalise the information. With regard to the cost of Microcapsulated Paper, Sena (2008) points out that "both in Brazil and throughout Latin America, this paper can only be found by import, which greatly increases its cost" (p. 107) (Appendix 5).

As with two-dimensional representations, the process of designing and making tactile representations in three dimensions is characterised by some specific techniques, which vary according to the scale of the representation. Two techniques stand out, one used to represent relief forms and the other, reduced spaces.

With regard to relief models, Sena (2001) emphasises the technique described by Simielli that:

> It consists of compiling contour lines from a topographic base, on tissue or tracing paper; transposing each curve onto Styrofoam plates, which are then superimposed and glued together; and finishing with plaster and/or putty to create the relief; finally, the model is sanded and painted for greater resistance (SENA, 2001, p. 32).

The representations resulting from this technique make accessible to visually impaired students structures and spatial organisations that are impossible to grasp through tactile exploration in the real world, which enhances the teaching-learning process about the dynamics and processes correlated with the represented forms.

Models of smaller spaces can be built using the collage technique. This begins by measuring the real space and/or analysing its floor plan, and after converting the measurements to the scale adopted, the model is built. Its elements are represented by different textures and colours.

The collage technique makes it possible to use different materials, which do not have to match those of the objects in reality. Ventorini (2012) describes the materials used in one of the models in his doctoral research:

> The blackboard was represented by a rectangular piece of felt and the windows and door were cut out of the cardboard box that represented the classroom. Cardboard boxes of different sizes were used to represent the furniture (the desks were represented by matchboxes, the teacher's desk and the cupboard by medicine boxes) (p. 109).

After the representations have been built, a new test is carried out, in which the cartographic products are analysed, preferably by visually impaired people who have more tactile training. In this respect, Almeida (2011) states that "map evaluation by visually impaired users (low vision or blind) is essential, since mappers are usually people who see and do not master tactile perception" (p. 151).

Given the diversity of materials, types of representation and techniques used, the definition of conventions in Tactile Cartography is difficult and does not reach unanimity among researchers and producers of materials. One of the aspects that contributes to this is the heterogeneity of users, as explained by Vasconcellos (1993):

> This group of users is extremely heterogeneous and complex, which makes it difficult to establish rules and define principles for making and using cartographic material. These principles are feasible to establish, but are

limited to certain conditions, such as prior training and user skills (p. 51-52).

Although some researchers advocate the adoption of conventions in Tactile Cartography, even drawing up standardised symbols for more general information, it is believed that diversity in representations is interesting and even more coherent with the inclusive perspective.

With the exception of maintaining symbols between cartographic representations in a collection, recognising the specificities of spatial perception and tactile graphic coding not only values the particularities of DV subjects, but also takes into account the potential and limitations of each reality. This facilitates the adoption and dissemination of tactile cartographic products in a wide variety of contexts, especially in schools.

2.3.2 Tactile cartography and geography teaching

In the process of understanding geographical knowledge, vision takes on great importance, as emphasised by Vasconcellos (1993):

> Visualisation is important, if not essential in human thinking; visual thinking penetrates all human activities, including abstract and theoretical ones, and it should be all the more present in Geography, both in the investigation of questions and in the presentation of results (p. 31).

In this context, cartographic knowledge is also built on visual information, which makes it essential to develop tactile cartographic products and teaching methodologies that contribute to the learning of visually impaired students.

In order to address this problem, Vasconcellos (1993) developed the Introduction to the Use of Maps Programme, which consists of a methodology for teaching geography to visually impaired students. This proposal was based on the author's research, studies and experiences, which highlighted the need to introduce basic concepts and cartography content before working with maps.

The basic concepts to be covered were graphic language, scale (proportion), location/orientation and points of view. To this end, the following activities were proposed:

> - tactile graphic language (introduction of tactile variables) activity: memory game on the flannelgraph with various sets of raised graphic variables, including a synthesis set.
> - scaling (reduction in size and distance) activity: carpet exercise, with the transition from reality to model, using a car and a person as an example; floor plan of the classroom.
> - points of view (horizontal and vertical) activity: representation of objects seen from the front and from above, excluding the notion of perspective (oblique) which depends fundamentally on vision.
> - location/orientation (coordinates and cardinal points) activity: "geographic battle" game with coordinate grid first with letters and numbers, then with the hemispheres (VASCONCELLOS, 1993, p. 43-44).

Vasconcellos (1993) used embossed graphic materials, exercises and games, which were planned and developed especially for the implementation of the proposal. Carmo (2009) points out that as well as contributing to the teaching of cartography, the Introduction to the Use of Maps Programme also aimed to arouse interest and motivate DV students to explore the "cartographic universe", which until then had been essentially visual.

It should be noted that Vasconcellos (1993) established the use of three-dimensional models and relief mock-ups as an important preceding step to working with maps, which, because they are less abstract representations, help visually impaired students to perceive and understand space.

Understanding that the tactile sense captures fragmented information, three-dimensional miniature representations enable a globalised perception of objects and environments, which is often not possible through tactile exploration of the real object/environment alone.

Furthermore, improving three-dimensional perception as a result of working with tactile models is essential for understanding the vertical vision used in floor plans and maps. Almeida and Nogueira (2009) state that "after this clarification [the concept of three-dimensional space] we can then introduce the concept of 'vertical vision' used in making conventional maps" (p. 112).

Among the topics covered in the Introduction to the Use of Maps programme, the concept of cartographic scale stands out. The aim of working with this concept was to introduce the notion of scale to children who did not know the decimal metric system and to facilitate understanding of this notion with those who already understood this system. Vasconcellos (1993) used as teaching resources: carpets and toys (cars and dolls) in various sizes, string and a model.

Vasconcellos (1993) reports on the activity with carpets, trolleys and dolls:

> Initially, a carpet cut to the actual size of a car (jeep) is rolled out and the students are arranged within the space delimited by it. From then on, they use toy jeeps that are 8, 55 and 160 times smaller than the real thing. To check the reduction, a string is used to measure the real jeep and it is folded until it is 8 times smaller (p. 91-92).

According to the author, by recognising that the larger rug corresponded to the size of the real car, the students were able to see the proportion between the two. The activity with the smaller rugs and toy cars enabled the students to understand the notion of reduction (VASCONCELLOS, 1993) (Figure 3).

Figure 3 - Materials for the scaling activity Source: Vasconcellos (1993)

The author also explains that the exercise on scale was applied in groups, without individual assessment. During the analysis of this activity, she considered that the students made good use of the concepts of reduction and proportion, as well as showing interest in the proposal (VASCONCELLOS, 1993).

It should be noted that in the work on Tactile Cartography carried out in Brazil, there is little reflection on the construction of the concept of cartographic scale with DV students. Professor Regina Vasconcellos' PhD thesis (1993) is considered to be the reference that presents the most systematised discussion of the subject.

Although the problematisation of the construction and understanding of the concept of cartographic scale is not vast, the question of its representation (as a peripheral element) in tactile cartographic products has been addressed more frequently. It is defined that the scale should be indicated by a raised line and appear at the bottom of the map. Sena (2008) has defended the effectiveness of the graphic scale in tactile cartography:

> With the graphic scale, the user can explore the map with both hands: one is fixed on the scale and the other is travelling the distances. In this way, they can establish a relationship of proportion between the line segment of the scale and the real distances represented (SENA, 2008 p. 115).

In this sense, cartographic scale is a significant element in the process of analysing and interpreting tactile cartographic products. By helping to establish relationships between the representation and real space, this concept consequently improves the perception and spatial understanding of visually impaired people. This context makes cartographic scale an important topic in Tactile Cartography, reinforcing the need to discuss and address it more systematically.

CHAPTER 3

METHODOLOGICAL STEPS: GUIDING AND BEING GUIDED ALONG THE WAY

This research was based on Participatory Research methodology, which is defined as a qualitative approach that presupposes the valorisation of practice in the research process, considering that "if a theory does not lead to practice, it was never even a theory, because it will be an unreal or alienating discourse, from another world" (DEMO, 1982, p. 59).

Considering that in Participatory Research, practice is indispensable to theory, the choice to carry out research of this nature required the definition of a field for the development of the empirical stages, and the Brazilian Centre for Rehabilitation and Support for the Visually Impaired (CEBRAV), located in the city of Goiânia (GO), was established as such.

This decision stemmed from the acceptance of CEBRAV's pedagogical coordinator and the Maths teacher at the institution to maintain the partnership established during the monograph research. In addition, the possibility of working with different students, with different types and levels of visual impairment, as well as the knowledge of CEBRAV's spatial organisation, acquired in the work mentioned above, reinforced the institution's suitability for the proposal.

On 9 April 2014, the first meeting was held with the supervising teacher to determine the students who would take part in the activity. Five students were nominated to attend the maths workshop on Wednesdays from 7.15am to 11.30am and on Thursdays from 3.30pm to 5.30pm. Monitoring the maths workshop on the days and at the times set made it possible to see that two of the five students indicated were not regular attendees. The other students in the suggested group were therefore chosen as the research subjects.

The stage of observing these subjects began. It took place mainly in the Maths workshop, and sporadically in other activities, such as Braille class and football. The observation of the students was guided by a script (Appendix 6) and systematised in a field notebook.

During the sessions, the students were informed of the observation and there was no interaction with them. At the end of each meeting, the aspects observed were discussed with the supervising teacher, in the absence of the students. The characteristics of the research subjects were also analysed by looking at their medical records, which is described in section 3.1.

The observation took place during the first half of 2014 and the month of August that same year. Together with the observation, Braille classes were held during this period, with the aim of improving the process of designing and making teaching resources and developing the activity. And Orientation and Mobility (OM) classes, in which it was possible to learn OM techniques and how to use a cane, as well as to experience, albeit minimally, experiences related to spatial perception, location and locomotion based on visual deprivation.

The readings and reflections of authors on tactile cartography signalled the need to learn more about the process of creating tactile cartographic representations. So, on 13, 14 and 15 May 2014, a technical visit was made to the Teaching and Learning Materials Laboratory (LEMAD) at the University of São Paulo and to the Laramara Institute in São Paulo (SP).

At LEMAD, under the supervision of Professor Waldirene[22] , it was possible to gain access to the laboratory's collection of tactile teaching resources, learn about various materials and techniques and make some representations. In

[22] Waldirene Ribeiro do Carmo: Technician in charge of LEMAD/USP.

addition, taking part in the workshop entitled "Tactile Cartography and Inclusive Geography Teaching", given by this teacher to teachers from the Mogi das Cruzes (SP) state network, made it possible to deepen discussions on the foundations and specificities of Tactile Cartography in the context of Geography teaching.

In the company of teachers Carla[23] and Waldirene, we visited the Laramara Institute to see its infrastructure, the specialised services on offer, the care areas and the resources, materials and programmes for working with visually impaired people. Professor Waldirene kindly provided specific materials for making tactile representations, as well as providing theoretical references on this subject and related topics.

At the same time as the observation stage and taking into account the experiences and knowledge resulting from the technical visit to LEMAD and the Laramara Institute, the choice of spaces to be represented was made. Thus, with the aim of addressing the concept of cartographic scale based on proportionality relationships in the students' lived space, the decision was made to represent two spaces in the CEBRAV building: the Maths workshop room and the first floor. These spaces are widely used by the students, who are partly familiar with their organisation and dynamics.

The Maths workshop room is accessed from inside another room, meaning it is not directly connected to the corridor. Measuring 6.26 metres long by 3.53 metres wide, the room consists of the following elements: window, door, bookshelf, cupboard, desks, chairs, air conditioning, fan, wastepaper basket, books and notebooks, educational games and various objects (Figures 4 and 5).

Figure 4 - Maths workshop room / CEBRAV Photo by the author

[23] Carla Cristina Reinaldo Gimenes de Sena: Assistant Professor at Universidade Estadual Paulista Júlio de Mesquita Filho

Figure 5 - Maths workshop room / CEBRAV Photo by the author

The first floor consists of a corridor 35.5 metres long and 2.70 metres wide. Along its length there are six doors on the right and five doors on the left (the staircase was used as a reference) (Figures 6 and 7). In the centre of the corridor there is a pillar, located between doors 4 and 5 (on the right) and 9 and 10 (on the left) (Figure 8), as well as a lift at the end (Figure 9). The corridor gives access to eleven rooms, seven of which have internal subdivisions.

Figure 6 - CEBRAV corridor Photo by the author

Figure 7 - Maths workshop room in the CEBRAV corridor Photo by the author

Figure 8 - Pilaster in the centre of the CEBRAV corridor
Photo by the author

Figure 9 - CEBRAV corridor lift Photo by the author

Once the spaces for the representation had been defined, the stages of designing and making the tactile teaching aids and planning the didactic sequence began, which are covered in topics 3.2 and 3.3 respectively.

3.1 The research subjects

As with conventional cartography, the process of designing and making tactile representations involves questions such as: What will be represented? Who is the material intended for? When will it be used? What results are expected? (CARMO, 2009). It is therefore important to know who the users are in order to build effective and meaningful representations.

The supervising teacher was fundamental in choosing the subjects, and after the initial observation and considering their attendance, it was decided to work with three students who are enrolled in the Goiás State Education Network and, among other activities at CEBRAV, attend the Maths workshop.

An important element in the choice of subjects was the fact that they were in grades in which, according to the Reference Curriculum of the Goiás State Education Network (2012), they were expected to work with the content of cartographic scale. This theme appears explicitly in the content and learning expectations for the 4th, 6th, 7th and 9th grades of primary school and the 1st [a] grade of secondary school.

The adoption of this criterion does not remove the concept that cartographic knowledge, including the concept of cartographic scale, permeates the other subjects of Geography and should therefore be considered in all years and grades of basic education.

In order to understand the research subjects, the following basic information was defined: age, type and duration of visual impairment, tactile training (mastery of Braille writing and reading, identification of different textures and materials), orientation and mobility skills, psychological aspects, schooling and length of time attending CEBRAV. This information was obtained by analysing medical records, talking to the supervising teacher and observing the students.

Student A, aged 14, has low vision caused by Optic Nerve Atrophy and his visual impairment has been established since birth. He is aware of his disability and his family's attitude is characterised by encouraging him to take part in activities and not overprotecting him.

Demonstrating a great ability to use the visual residue, student A sees colours, textures and shapes, especially in bright environments. His reading and writing process is based on writing in ink, without the need for significant magnification of letters, numbers and symbols. It should be noted that the only optical resource used by this student is prescription glasses.

The exploration of objects and environments is carried out through visual perception, with little use of the tactile sense. Their orientation and mobility skills are developed and the student gets around without difficulty, even using public transport on their own.

Student A is in the 8th year of primary school and has been attending CEBRAV for 3 years, currently taking part in Maths and Portuguese workshops and Psychology sessions. Although he is shy and introverted, he is affectionate and polite towards the supervising teacher. He is also very interested and involved in carrying out the activities proposed in the Maths workshop.

Student B is 19 years old and blind. His process of blindness, caused by Retinopathy of Prematurity, was completed in the first few years of his life. Like student A, student B is aware of his disability and is not embarrassed to talk about it. His family relationship is characterised by an absence of over-protection or attitudes that limit his development.

Student B has a good improvement in his tactile sense, being able to perceive, differentiate and identify different textures, shapes and sizes, which enhances his process of exploring and understanding objects and environments. He is literate in Braille, showing speed and confidence in reading and writing in this system, as well as using the soroban correctly .[24]

Demonstrating significant development of his orientation and mobility skills, student B uses a cane and OM techniques to get around. It should be noted that this student lives in a different municipality to CEBRAV and travels home - CEBRAV and CEBRAV - home by public transport alone.

Student B is in the second year of secondary school and has been attending CEBRAV for three years. He currently takes part in the following activities: maths and Portuguese workshops, IT, music, orientation and mobility and football. He is friendly, affectionate and outgoing with his classmates and teachers, and is always interested and participative in maths activities.

Student C, 19, has blindness in her right eye and low vision in her left eye. Her visual impairment is the result of

[24] Soroban is a calculating instrument made up of rods and beads, which enables visually impaired people to perform mathematical operations.

congenital toxoplasmosis and she suffered this condition a year and a half ago. The student recognises her visual impairment, but is reluctant to talk about it. With regard to her family, she feels supported by her parents and brother, but excluded by her uncles, grandparents and cousins.

She also attributes the reason why she is discriminated against at school to her visual impairment. Student A often described embarrassing situations she experienced in the school environment, especially in relation to the school management and teachers. These conflicts meant that even though she was enrolled, she didn't attend classes for part of the first semester. In one of her speeches, she said: "I feel bad at school, because the lessons take place as if I didn't exist".

Student C's visual residue is limited to the perception of light, and it is through tactile perception that her exploration process takes place. In this way, she perceives and differentiates textures, shapes and sizes, identifying objects and environments. She started Braille literacy in the first half of 2014 and although she can read and write in this system, she is still confused and unsure about some symbols and rules. She still doesn't use the soroban.

Her orientation and mobility skills are limited. The student does not use a cane and does not attend OM classes. She gets around indoors (the CEBRAV building) by tracking, and outdoors she needs the company of sighted people.

Student C is in the second year of secondary school and has been attending CEBRAV for two years. She currently takes part in the following activities: Maths and Portuguese workshops, IT and Braille literacy. During the Maths sessions, it was possible to see that the student is insecure about doing the activities and when she makes a mistake, she feels guilty about what she has done. Although she is withdrawn, she is polite and gentle with her teacher, but has little interaction with her classmates.

3.2 Designing and making tactile teaching aids

The process of designing and making the tactile teaching aids began by defining the spaces to be represented, and was organised into three stages. The first was characterised by evaluating the materials that could be used; the second was the construction of the representations themselves, including defining the scales, converting measurements, measuring and cutting out the textures and assembling and finishing the representations; and the third was the evaluation of the representations.

3.2.1 Evaluation of materials - Sensory Notebook

In the process of building tactile cartographic representations, the sensory exploration and analysis capacity of the individual for whom this resource is intended must be taken into account. It is therefore essential to know the tactile training of blind people, as well as the association of this sense with the visual residue of people with low vision. According to Vasconcellos (1993, p. 49):

> The effective use of visual and tactile graphic language depends on effective and reasonably long training. Sensory stimulation should start very early, before the child enters the first grade of primary school. Children need to be very well prepared to understand and use the language of maps, which is all the more necessary in the case of visually impaired pupils.

The influence of sensory perception on the process of interpreting, analysing and understanding tactile cartographic representations means that evaluating the materials to be used, as well as the representations themselves, are essential methodological steps in the process of building these resources. In this respect, Carmo (2009) states that "user evaluations of the material are fundamental to producing effective maps" (p. 60).

In this research, as contact with the participants was restricted to the time of the investigation, the testing stage became even more important. Thus, based on the model presented by Vasconcellos (1993), a Sensory Notebook was built comprising 56 materials potentially suitable for making teaching resources.

In addition to the explanatory text, the Sensory Notebook is made up of a numerically organised list of materials and sample textures. Each page contains eight textures, spaced apart (Figures 10 and 11).

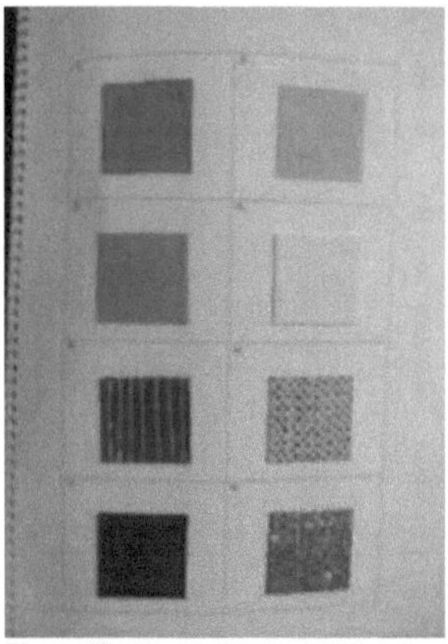

Figure 10 - Page 8 of Volume II Photo by the author

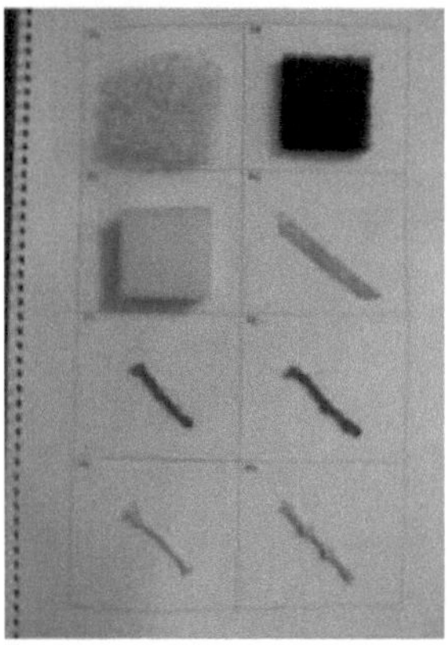

Figure 11 - Page 12 of Volume II Photo by the author

The materials were assessed individually with each student, taking into account their characteristics and specificities. After a personal introduction, an explanation of the aims of the activity, in which it was stressed that the evaluation was focused on the material and not the students, and a description of the organisation of the notebook, the students analysed the

textures. During the activity, we took into account both spontaneous speech and the answers given based on specific questions about the materials.

On 3 September 2014, an assessment was carried out with student A. This activity was structured around the following objectives: checking whether the student could see colour; assessing his tactile perception; discussing the different colours and textures of materials; defining which textures were pleasant and unpleasant; and checking the identification of materials used in everyday life. At first, the materials were only observed through visual perception, which was combined with tactile perception in the second stage of exploration.

On 18 September 2014, an assessment was carried out with student B based on tactile exploration. The aim was to assess the student's tactile perception; discuss the different textures; define which textures were pleasant and unpleasant; and check the identification of materials used in everyday life. The assessment with student C took place on 14 October 2014, based on tactile perception and considering the same objectives as student B.

It should be noted that the arrangement of the materials in the Sensory Notebook was adequate for the analysis, and the students showed no difficulties in locating the textures, exploring them individually and interrelating them. Although they said they had never taken part in a similar activity, all the students understood the proposal and did not feel uncomfortable doing it.

The three students considered most of the textures to be easy to perceive and pleasant in terms of tactile exploration. In this way, student A judged that only one texture was not easily perceived and that three textures were uncomfortable to the touch. He also pointed out that seven materials had colours that were difficult to see.

Student B, on the other hand, considered that of the 56 textures presented in the Sensory Notebook, two were difficult to perceive and only one was unpleasant to the touch. Student C rated ten textures as difficult to perceive and one as uncomfortable.

Material 30 (sandpaper) was considered uncomfortable by students A and C, while material 33 (vegetable loofah) was singled out by students A and B on this same criterion. Regarding materials with barely perceptible textures, students A and C agreed on material 25 (crepe paper - black) and although student C characterised 10 materials as uncomfortable, none of these coincided with those pointed out by student B.

It's interesting to note that none of the materials considered by the students to be difficult to perceive, unpleasant to the touch, or of an unfavourable colour, were used in the representations constructed.

3.2.2 Construction of tactile teaching aids

In addition to the central objective of the research and the evaluation of the materials, the definition of the type of representation was influenced by the students' accounts of their experiences with models. Student A said he was familiar with this resource, which had already been worked on in science. Students B and C, on the other hand, said they were unaware of these representations, which are not part of their Geography lessons.

The decision was then made to build models using the texture collage technique. As three-dimensional representations, the models were thought to encourage both the perception of proportionality relationships and make a more significant contribution to understanding spatial organisation.

The first model built was of the Maths workshop room (Figure 12). After measuring this space using a 50 metre tape measure and taking photographs, a scale of 1:25 was defined. Among the elements that make up the room, we chose to represent: door, window, bookcase, cupboard, two blocks of desks (each block is made up of six desks), air conditioning, fan and curtain.

Figure 12 - Mock-up of the maths workshop room Photo by the author

The model was built in a shoe box, which was cut out to fit the measurements determined by the 1:25 scale (Figure 13). All the elements represented obeyed the proportion, except for the fan due to the difficulty of measuring this object in the real environment.

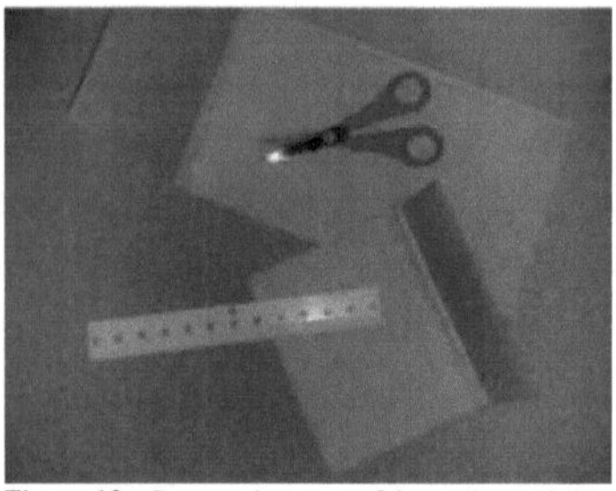

Figure 13 - Construction stage of the maths workshop room model Photo by the author

Once the measurements were appropriate, the representation was built using different textures and colours, selected after evaluating the materials. The shapes and textures of the representations were as close as possible to those of the corresponding real elements.

So, in order to simulate wood, the representation of the bookcase was built with ice-cream sticks and bamboo skewers and the representation of the curtain was made with TNT, which was the same material as this object in reality. In the representation, the elements were arranged according to the organisation of the room in reality and fixed with glue.

Table 5 shows the materials used to represent the maths workshop room and the objects that make it up:

Objects	Materials
Walls	0.2 mm cardboard painted with white paint
Floor	Cork
Door	Green EVA glitter
Window	Blue suede paper
Bookcase	Cardboard, popsicle sticks and bamboo skewers
Cupboard	Black cardboard and accordion paper
Wallet block	Cardboard, red graphite EVA, bamboo skewer and straw
Air conditioning	Cardboard and smooth EVA in yellow
Fan	Smooth orange EVA textured with blue coloured glue
Curtains	Purple TNT

Chart 5 - Materials used in the maths workshop room model Source: Prepared by the author.

The legend was built along with the model using cardboard and drawing paper. This contained the title of the representation, the symbols, preserving the shapes and materials and their meanings in enlarged ink (Verdana 22) and

Braille, and the graphic scale, represented by a 1 centimetre maquette, also written in both ink and Braille (Figures 14 and 15).

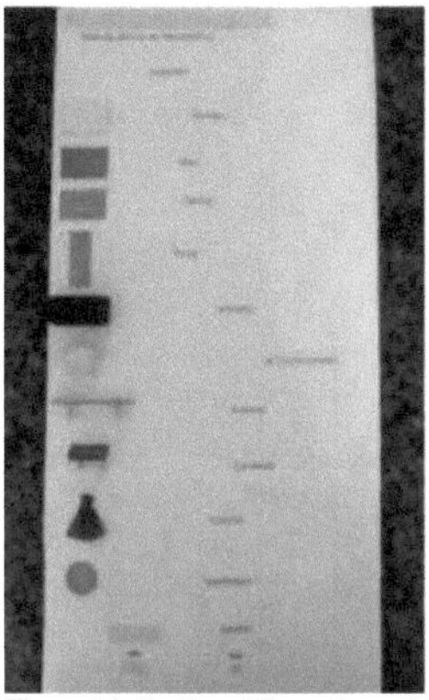

Figure 14- Caption of the maths workshop room model Photo by the author

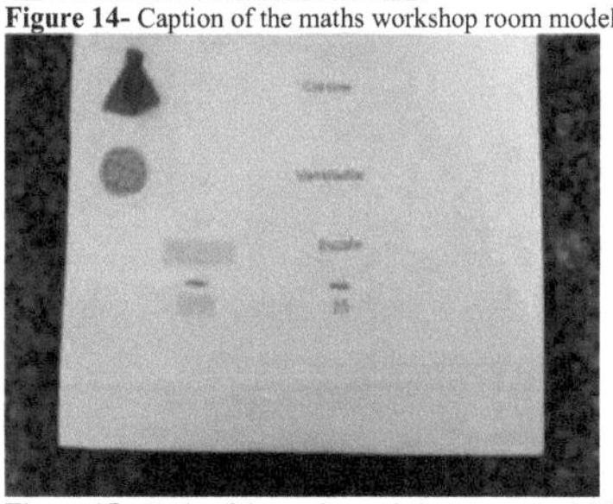

Figure 15 - Scale of the maths workshop room model Photo by the author

The mock-up of the corridor was built on the basis of the building's floor plan, provided by CEBRAV's director, and photographs. As the aim was to prioritise the physical structure of this space, the following elements were represented: corridor, rooms, lift, walls, floor, doors and windows. The scale used was 1:50, from which the measurements were converted. The model was 71 cm long and 30 cm wide (Figure 16).

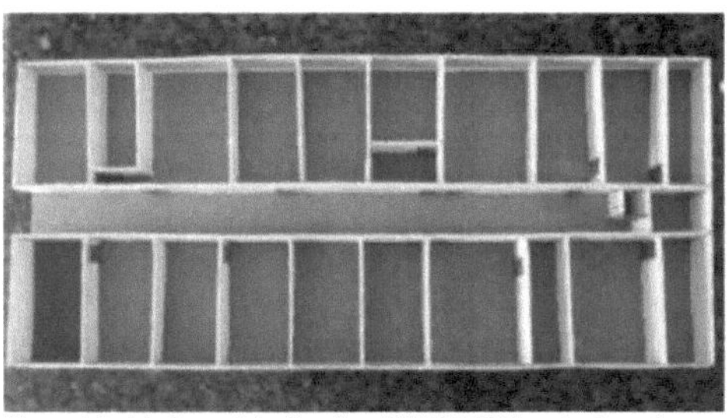

Figure 16 - Mock-up of the first floor / CEBRAV Photo by the author

The height of the doors and the width of the windows in the representation were defined using an empirical scale. As the height of the walls in the model was 5.5 cm, if the scale was followed, the doors would be 5.2 cm, which could make interpretation difficult, so the doors were represented as 4.5 cm high. In the case of the windows, this decision is justified by the fact that they are not represented on the building's plan, and it would be impossible to measure them in reality, given the number of rooms on the first floor.

During the construction process, the decision was made to keep the materials used in the previous model. So the walls, floor, doors and windows were built with the same textures in both representations. Only the floor of the Maths workshop room in the corridor model differed from the standard, due to the need to emphasise this space for carrying out some of the activities planned in the didactic sequence.

To facilitate analysis and interpretation of the first floor model, the corridor doors, including the lift, were numbered from 1 to 12 (Figure 17). The numbers were fixed above the doors and written in ink (Verdana 22) and Braille, and a text was drawn up (Appendix 7), also in the two forms of writing, in which the numbering was related to the room names.

Figure 17 - Door numbering on the first floor model / CEBRAV Photo by the author

As with the model of the Maths workshop room, a legend was also made for the representation of the first floor, explaining the meanings of the textures and shapes, the title and the graphic scale (Figure 18).

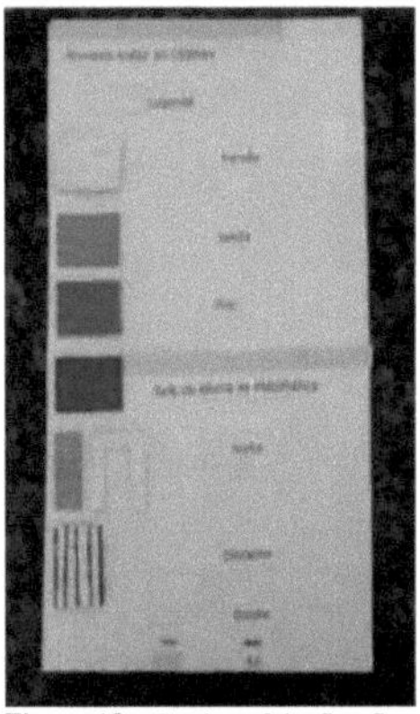

Figure 18 - Caption of the first floor model / CEBRAV
Photo by the author

In addition to the two tactile models, we decided to represent some objects, the criterion for choice being the existence of these in the Maths workshop room and on the first floor. We then represented a book on a scale of 1:1, a chair on a scale of 1:10, a lollipop on a scale of 3:1 (Figures 19, 20 and 21) and the block of desks, the cupboard (objects in the Maths workshop room) and the pillar in the corridor on a scale of 1:50 (Figure 22).

Figure 19 - Representation of the book

Photo by the author

Figure 20 - Representation of the chair
Photo by the author

Figure 21 - Representation of the lollipop
Photo by the author

Figure 22 - Representation of the block of desks, the cupboard and the pilaster Photo by the author

It should be noted that the materials used to represent the objects were part of the sensory notebook. The block of desks and the cupboard, on a scale of 1:50, were built similar to their counterparts in the model of the Maths workshop room.

3.2.3 Evaluation of Cartographic Representations

Given that the models were not built simultaneously, these representations were also assessed at different times. The criteria used were: correspondence between the representation and reality; adequacy of shape, size and materials used and identification of textures considered unpleasant.

The first representation submitted for assessment was that of the Maths workshop room. This activity was carried out with CEBRAV's braille teacher, who is blind and knew the organisation of the space represented. The assessor considered that the model was similar to the real environment in terms of the shape of the room and the elements; the arrangement and organisation of the objects in the space; and the distances established between the structure of the room (wall, door, window) and the objects and between the objects themselves.

The size of the representation was considered adequate for tactile exploration, as was the size of the objects. The textures were characterised as perceptible and easy to differentiate, as well as being pleasant to the touch. During the exploration of the cupboard, which was represented using cardboard and accordion paper, the teacher said: "This is the book cupboard. You've added a texture that really resembles the organisation of books and notebooks". However, when this representation was constructed, the relationship between texture and the object's function was not established. This association was used in the activity with the students, especially at the stage of understanding the elements represented.

The evaluation of the second model was carried out by the supervising teacher (sighted) and a CEBRAV employee (sighted), who had detailed knowledge of the organisation of the rooms on the first floor and their internal configurations. As the textures were maintained in both representations, their suitability for tactile exploration was inferred.

The assessors considered that the representation was similar to the real environment. Due to the length of the model (71 cm) and the fact that the students didn't know the first floor in its entirety, the teacher pointed out that the process of exploring this representation required more monitoring than the first one.

The CEBRAV employee proposed two changes to the model. The first was to correct the location of one of the internal doors in the visual rehabilitation room, and the second was to subdivide the computer room by adding a wall and a door. This information did not appear on the building's floor plan, as it was the result of a refurbishment. So after readjustment, the model was analysed again by the employee and then approved for the activity with the students.

3.3 Designing the didactic sequence

The consolidation of a teaching-learning process that enables reflection and understanding of themes and concepts and that is not limited to the simple transmission and memorisation of definitions depends on how this process is organised, which makes the Didactic Sequence of great importance. According to Reorientação Curricular: currículo em debate - Goiás (2009):

> A planned teaching and learning situation, organised step by step and guided by the objective of promoting defined learning. These are sequenced activities, with the intention of offering challenges of varying complexity so that students can gradually appropriate knowledge, attitudes and values that are considered fundamental (p. 13).

The Didactic Sequence provides for the configuration of procedural teaching, in which the objectives to be achieved are the basis for structuring the planning, the stages of application and the evaluation process, these three moments being interdependent and non-linear (ZABALA, 1998).

The didactic sequence that guided the activities planned in this research was based on the model presented by

Reorientação Curricular: currículo em debate - Goiás (2009), which has the following structure: presentation of the work proposal; survey of students' previous knowledge; expansion of knowledge; systematisation of knowledge and evaluation.

The presentation of the work proposal consists of explaining and sharing the general process that will be developed. Gathering prior knowledge is carried out by problematising the topics that will be covered, in order to check what the student knows about these issues, and can be developed through "conversation circles, reading images and/or written texts, problem solving, debates, among other strategies" (GOIÁS, 2009, p. 14).

Considering the statements and answers obtained in the survey of prior knowledge, there is the stage of expanding knowledge, in which the scientific aspects of the topic studied are reflected on and discussed. The aim is to broaden and deepen the students' understanding.

The systematisation of knowledge, in turn, consists of "retracing the path, organising the main notions and concepts worked on" (GOIÁS, 2009, p. 14). And assessment is procedural and is established through reflections and records by the teacher and the student regarding consolidated learning.

3.3.1 Teaching sequence

Planned time: 3 meetings of 40 minutes each.

Objectives:

- Discuss the concept of representation;
- To help in the process of interpreting and understanding tactile representations;
- Promote students' contact with tactile representations, prioritising proportional relationships and the concept of cartographic scale;
- Contribute to understanding the notion of proportion and the concept of cartographic scale;
- To assess whether the development of the notion of proportion, based on lived space, helps in the understanding of spatial organisation and, consequently, in the improvement of orientation and mobility skills.

Materials used: tactile models, various objects and their representations, different sizes of string and A4 paper.

Presentation of the proposal

Briefly return to the activity of evaluating the materials and then explain the Didactic Sequence proposal, highlighting the planned activities, the didactic resources that will be used and the objectives to be achieved. At this point, it is important to emphasise that the focus of the evaluation and analysis is the teaching resources and not the students.

Gathering prior knowledge

Through a conversation, students will be asked to talk about their experiences in relation to Geography class. After this more general context, they will be asked specific questions about cartography. Firstly, they will be asked if they have or have ever had contact with models in Geography lessons, or in other contexts. Then the following questions will be asked: What is a representation? Do you know what proportion is? Do you know what scale is? During the conversation, both answers based on school experiences and those from everyday life will be encouraged and valued. Students will then be asked to explore the representation of the chair, from which the discussion about representation will begin.

Broadening knowledge

Concept of representation

Considering the students' thoughts on the representation of the chair, Le Sann's (2007) definition will be presented[25] , and this definition will be analysed by associating its aspects with the representation explored by the students.

Types of representation

To discuss the types of representations, the element of size (distance) will be considered, since this aspect is present in the notion of proportion and, consequently, in the concept of cartographic scale. Students will then be asked to explore the representations of the book, the lollipop and the chair and compare them with the corresponding real objects. This activity will make it possible to understand the existence of representations that are larger, equal and smaller than the objects represented.

Interpreting the model of the Maths workshop room

This stage will begin with questions such as:

- Do you know the Maths room?
- Is the room big or small?
- What are the elements that make it up?
- How are these elements organised?

Through the perception of the size of an A4 sheet of paper, the discussion will be based on the following proposition: What would have to be done to represent the Maths workshop room on this sheet of paper? In view of the need to reduce the size of the Maths workshop room, a model of this space will be presented. The students will first freely recognise the representation and then interpret it based on the legend. Confirmation of the elements represented in this model, as well as their arrangement in the space, will take place by exploring the real environment.

Concept of proportion

In order to help students understand that the notion of proportion is maintained between the parts of a representation and between its various elements, they will be asked to compare the parts that make up the representation of the chair with those of the real chair. Thus, based on the assumption that the representation is smaller than the real object, the student will see that the backrest, seat and legs of the represented chair are smaller than their real counterparts. Afterwards, will ask the student to explore each object represented in the model of the room, analysing it in terms of itself, its relationship with the model as a whole and, finally, its correspondent in reality. Then it will be explained that between *the* representation and the object/space represented there is a relationship of size called proportion, which is maintained between the constituent parts of a represented object (example, the chair) and between all the elements of a representation (example, the model of the Maths workshop room).

[25] According to Le Sann (2007), representation consists of reconstituting a particular object or space after observing and perceiving it. Thus, most of the time the representation has similarities with the object/space it represents, but it is not that object/space.

Concept of scale

Emphasising that the reduction of the model of the maths workshop room in relation to the real environment is not done randomly, the graphic scale of this representation will be shown. By exploring the scale, the student will realise that the size of the line in relief coincides with the tip of his finger and that it actually corresponds to 25 sizes of the one he explored, so a 25 cm string will be used. After explaining that the size ratio is 1:25, questions will be asked such as: Part of the block of desks on the model (measuring one fingertip) corresponds *to* how many parts on the real card block? A part of the door on the model (measuring *a* fingertip) corresponds to how many parts of the real door? etc. The concept of scale will then be explained, inserting the notion of centimetres, i.e. one cm on the model corresponds to 25 cm on the real thing.

Interpretation of the first floor model

The model of the first floor will first be explored freely and then with the help of the legend, after which the spatial organisation of the space represented will be discussed. To do this, the student will be given a text containing the number and name of the rooms and the analysis will be carried out emphasising the location of these rooms in the corridor, as well as their internal configurations. Returning to the concept of scale, it will be explained that the model was built on a scale of 1:50, again using string to make this measurement clear. The concept of centimetres will then be reinforced and attempts will be made to calculate the real measurements *from* the model and its scale.

Comparing the models

Students will be asked to compare the two representations, considering aspects such as: size of the space represented, size of the representation, number of places represented, level of detail, number and size of objects represented.

Systematising knowledge

In order to systematise the knowledge built up, the stages will be recalled in the chronological order in which they were developed. Thus, when the activity is mentioned, the student will be asked to talk about what they learnt in each of these moments. The exploration of tactile representations will be revisited whenever necessary, as will the explanations of definitions and concepts.

Evaluation

The evaluation will be procedural and will be based on the students' responses and speeches. To this end, all the meetings will be audio-recorded and the speeches will be analysed afterwards. There will also be a number of activities. The first activity consists of experimenting with representations of various objects (at different scales) in each of the models. Considering that the objects exist in the real environment, this activity will also make it possible to address aspects of spatial organisation. The aim is to

- Compare the size of the objects with the size of the model;
- Analyse the proportional relationships between the representations;
- Reinforce ; understanding of the spatial organisation of the represented environment;

- Understand the influence of scale on representations.

The second activity will be based on comparing the models and their scales, starting with the following question: If we were to represent the entire CEBARV building on a scale of 1:100, what would this representation look like in relation to:

- Size of the represented space;
- Size of representation;
- Number of seats represented;
- Level of detail;
- Quantity and size of the objects represented.

CHAPTER 4

DIDACTIC PROPOSAL FOR BUILDING THE NOTION OF PROPORTION AND CARTOGRAPHIC SCALE WITH VISUALLY IMPAIRED STUDENTS

After the observation period and the evaluation of the materials, the pedagogical work began, which was planned as a Didactic Sequence. However, the responses obtained during the course of the activity influenced its structure, so that although it was based on the same didactic sequence, the execution of the proposal differed in some respects with each of the subjects.

It should be noted, for example, that the need for some materials was recognised during the course of the activity. At first, it was considered sufficient to only work with reduced representations, and so a representation of a chair (1:10) was built. It was only after the first meeting with students A and B that the need to work with representations of the same size and larger than the object represented was realised.

Given that the activities had to be carried out during the Maths workshop and that the research subjects attended it on different days and at different times, we worked individually with each student. Students A and B took part in three meetings and student C, due to repeated absences, in two, with each meeting lasting approximately 40 minutes. Recognising the particularities and differences between the activities, which may be the result of the also different profiles of the subjects (Chart 6), we chose to describe the pedagogical work with each student, with the results and discussions being carried out together.

Students	Age	Type of disability	Education	Time spent at CEBRAV
A	14 years old	Low vision	Year 8	3 years
B	19 years old	Blindness	2ª series	3 years
C	19 years old	Blindness o.d. Low vision o.e.[26]	2ª series	2 years

Chart 6 - Characteristics of the research subjects Source: Prepared by the author

4.1 Student A

The activities with student A took place on 5, 12 and 19 November 2014, in the morning, in the Maths workshop room. After a quick presentation, the structure of the activity, the number of meetings, the materials used and the objectives were explained. At this point it was emphasised that the focus of the assessment was the models and not the student.

In order to check his prior knowledge, student A was asked if he knew any models and if he had ever built or worked with such a resource at school. The answer was positive, and this work was carried out in Science. Turning to the context of Geography lessons, he was asked about representation, proportion and scale, which he said he didn't know.

After explaining the definition of representation, using a notebook as an example and the possibility of drawing it, the student explored the representation of the chair, from which the idea of reduction was approached. Priority was given to understanding that representation is not the real object, and this aspect was addressed through questions such as: "Could we sit on this chair?"; "Is this chair capable of supporting our weight?".

This led to a discussion about the organisation of the Maths workshop room, which the student said he knew

[26] The abbreviations o.d. and o.e. stand for right eye and left eye, respectively.

because he had studied there for a long time. When asked about the objects that make up the room, he pointed out: tables, a bookcase, a door, a window, a cupboard and chairs. When asked if he knew about the air conditioning and the fan, he immediately replied yes, pointing to them with his finger. Also at this stage, we tried to discuss the arrangement of the elements, emphasising their proximity, order and continuity in the space. Given student A's visual ability, tactile exploration of the real environment was not necessary.

In order to introduce the notion of proportionality, considering the size of an A4 sheet of paper (which was given to the student), it was hypothetically suggested that the student draw the room, asking what would be needed to carry out this task. At first the student was confused, but after defining that the room was large and the sheet small, he came to the conclusion that it was necessary to reduce this space in the drawing.

Once again, the student was asked to explore the representation of the chair, instigating the perception of the size relationship between the parts of the representation and these with their correspondents in reality. Thus, the proportion was called the size relationship established between two variables, in this case between the representation and the object represented.

The model of the maths classroom was presented to the student and before any activity, his ability to see the elements, shapes and colours of this representation was checked. The student explored the model visually, identifying all the elements with the help of the legend. At this point, we tried to link the organisation of the model to the information obtained from the (visual) analysis of the real environment.

After that, the exploration was based on the size variable, which was first observed individually in some elements, then in relation to the model and finally associated with their real counterparts. Finally, it was explained that the reduction and the resulting ratio are not random, but are established by a scale. After explaining this definition, the scale of the maths room was presented, which was 1:25. The meaning of this expression (1:25) was worked out using a 1 cm relief line segment and a 25 cm string.

At this point, the explanation of the scale was not based on exact measurements, in other words, it was not considered in centimetres. So the discussion went as follows: "a little piece, a fingertip, in the model corresponds to 25 pieces of this in reality".

The second meeting began with a discussion of the activity carried out on 5 October 2014, in which the student highlighted his contact with the model of the room. Going back to the definition of representation, it was described as "making what you have in one place into a smaller object".

In order to work on the types of representation, it was pointed out that in relation to size, there are three possibilities: smaller, equal and larger than the objects represented. To this end, we used the representations of a chair (already known to the student), a book and a lollipop (Figure 23), respectively, and explored these in articulation and comparison with the objects in reality.

Figure 23 - Visual exploration of the representations of the book, lollipop and chair, by student A
Photo by the author

When asked if the representations were real objects, the student promptly replied that they were not, demonstrating his understanding of this aspect. Once again, work with the model was proposed, based on the concept of scale, seeking to clarify that just like the notion of proportion, the scale relationship applies to all the elements of the model.

In this way, they were encouraged to think that a finger tip on the door of the model corresponded to 25 pieces on the real door, a finger tip on the table corresponded to 25 pieces on the real table, and so on. While carrying out this exercise, the student sometimes noted the size of the embossed line segment and the string.

Positioning the model of the first floor in the same direction as reality, the work with this representation began with free (visual) exploration, from which we proposed identifying the space represented (Figure 24). Student A quickly recognised that it was part of the CEBRAV building and used the legend to interpret the model.

Figure 24 - Visual exploration of the first floor model by student A Photo by the author

In addition, the spatial organisation of this representation was analysed using the text, which contained the number and name of each room in the corridor. First, the students were asked which rooms they attended at CEBRAV (the Maths room, the Portuguese room and the Psychology room) and asked to locate them. With the help of the text, all the other rooms were identified, and their internal organisations were also discussed.

The next step was to work with the scale of the model of the first floor (1:50). To do this, they explained again the meaning of this relationship with the line segment (a fingertip) and the string (50 cm). It was then defined that the line segment measured 1 cm and the corresponding string, 50 cm. The student showed that he knew the measurement in centimetres, extending this reasoning to the model in the Maths classroom.

Considering that the scales influenced the characteristics of the representations, a comparative analysis between them was coordinated. Thus, aspects such as: size of the space represented; total size and size by parts of the representation; number of places represented; level of detail and number and size of objects were analysed. At this point, student A had some difficulties, especially in relation to the number of places represented.

The following problem was therefore posed: if we placed one finger on the wall of the model, that size in real life would be 50 cm and if we placed two fingers, how big would they be? The student replied: 100 cm. Then he asked himself: to get 100 cm, in reality, in relation to the Maths room model, how many fingers would we have to place? Four fingers, said the student.

At this point, they tried to calculate the measurements of the room using the model, but this activity was hampered by the shape of the representation, which interfered with the positioning of the hands, the movement of the fingers and, consequently, the execution of the calculations.

The third and final meeting began with questions about representation, types of representation and proportion, of which only the last topic was not answered correctly. It was explained that proportion was a relationship of size between the representation and real space, affecting the representation as a whole, its elements and the constituent parts of these elements. The model of the Maths workshop room was used for this purpose.

Afterwards, the scales of the models were taken up again, using the decimal system. The student showed that these measurements applied to all the elements, which made it possible to infer his understanding of the proportional relationship. The activity was then carried out with different representations (chair, book, lollipop, table, cupboard and pillar) at different scales.

First using the model of the Maths workshop room (Figure 25) and then the first floor, the student was asked to include elements that are part of these real spaces. Without difficulty, the student realised the adequacy/inadequacy of the elements in relation to the models, associating this context with the differences in scale. It's interesting to note that in addition to scale/proportion, this activity addressed issues of spatial organisation, enabling the student to mentally articulate the representation and its corresponding real environment.

Figure 25 - Activity with representations on different scales, by student A Photo by the author

In order to reinforce the understanding that representations differ according to the scale adopted, the following hypothetical situation was proposed: we are going to build a representation of the entire CEBRAV building on a scale of 1:100 (the sizes were visualised by a 1 cm line segment and a 100 cm string). They were then asked about the size and quantity of the spaces represented, the level of detail and the size and quantity of the elements. The student answered all the questions correctly.

The steps of the activity were recalled, prioritising the definitions and concepts discussed during the activity. The student showed that he had understood the notion of proportion and scale. When asked if it was possible to estimate the real size of the space represented by the scale, student A said yes.

The activity ended with an evaluation of the whole process by the student, who emphasised, above all, the mock-ups, which were considered "good" because they made it possible to "get to know a lot of spaces".

4.2 Student B

The activities with student B were also carried out in three meetings, held on 4, 11 and 18 November 2014, in the afternoon, in the Maths workshop room. The first meeting began by presenting the structure of the activity, the expected number of days, the materials and the objectives. It was emphasised that the assessment was centred on the models.

Previous knowledge was checked by asking questions: Have you ever built or worked with models? Do you know what a representation is? Do you know what

proportion and scale? The student replied that he didn't know any models and couldn't remember any of these "names", and couldn't define whether these topics had been worked on in Geography lessons.

We then explained what representation is, using the chair the student was sitting on as an example. This was explored and compared to its representation, and the student was surprised and enthusiastic about the miniature. It was emphasised that a representation is not the real object, although it often resembles it.

When asked about his understanding of the Maths room, student B said that he knew some of its objects, including: the desk, the Braille machine, the bookshelf, the window, the door and the air conditioning, but he was confused about their spatial arrangement. Defining the room as large and realising the size of the A4 sheet, student B concluded that it was necessary to reduce the size of the room in order to represent it.

Based on the discussion about reduced representation, the student again explored the representation of the chair, now emphasising the size relationship between the representation and the real object, which was called proportion.

The model of the room was presented to the student, who explored it freely. During this stage, he showed great interest in interpreting the representation, but when he couldn't understand the texture of the fan, student B became frustrated. At this point, he considered it necessary to work with the legend, which made it possible to identify all the elements represented (Figure 26).

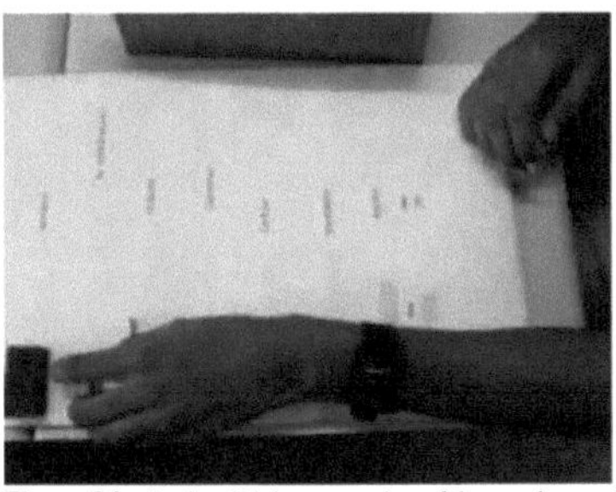

Figure 26 - Student B's interpretation of the caption on the Maths room model Photo by the author

Based on the student's interest in analysing the organisation of the Maths workshop room, he was asked to highlight the elements that were not represented in the model, of which he pointed out: chairs, books and a jar of sweets. With the aim of working on

order of the elements, the model was explored first and then the real environment. During this exploration, student B verbalised the arrangement of the objects in the room beforehand.

This stage made it possible to understand the relationship of proportion between the model, its elements and reality. Thus:

> Researcher: Let's compare the elements with each other. For example, is the door bigger than the air conditioning unit in the model?
> Student B: No (the student explores the air conditioning again). Yes, the door is bigger.
> Researcher: And if we compare the fan with the bookcase, which one is bigger? Student B: The bookcase.
> Researcher: In the real room, is the bookcase bigger than the fan?
> Student B: Yes.
> Researcher: And the door is bigger than the air conditioning?
> Student B: Also
> Researcher: This means that the elements in the model have the same proportions as in reality, in other words, if in reality the air conditioning unit is smaller than the door, in the model the air conditioning unit will be smaller than the door. If the fan is smaller than the bookcase in reality, will it be smaller or larger in the model?
> Student B: Smaller than the bookcase.

The second meeting began with a conversation about the first day's activities, in which the student emphasised working with the size of the objects and exploring the model. When asked if the representation was the real object, he promptly replied that it was not. In order to address the types of representation in relation to the size variable (smaller, equal and larger), the student explored the representations of the chair (Figure 27), the book and the lollipop, comparing them with their correspondents in reality.

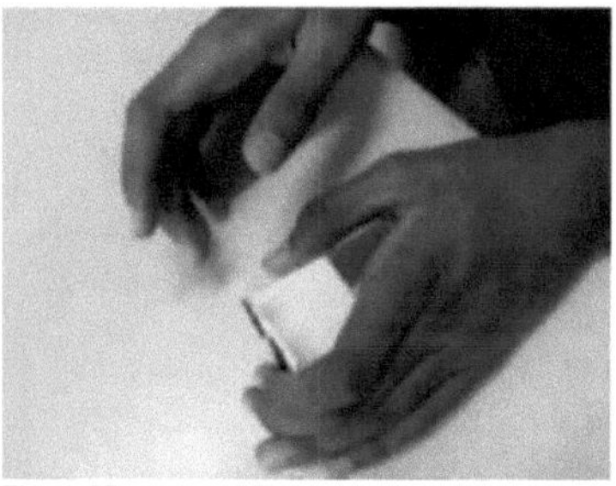

Figure 27 - Exploring the representation of the chair, by student B Photo by the author

The idea that the proportion affects all the elements of the model was taken up again, emphasising that its reduction in relation to reality is not random, thus introducing the concept of scale. The student explored the line segment, from which it was established that that size in the model (a fingertip) corresponded to 25 sizes of that in reality (25 cm string). Finally, it was explained that this size relationship also applies to all the elements of the representation.

The student had difficulty understanding the size of 25cm, and explored the string several times. This also hindered his understanding of the size relationship between the model and the real thing. It was clear that the student understood the reasoning, but that the notion of the measure itself was confusing.

Afterwards, the second model was explored freely and without identification. During this stage, student B was very focussed and after a while he said "it's representing the school here". Due to the repetition of textures referring to the same elements in the two models, the student interpreted the representation of the first floor, without using the legend.

While looking at the model of the first floor, the student questioned the internal organisation of some of the rooms. Thus, using the explanatory text, student B identified the order of the doors in the corridor and analysed the internal organisation of the rooms, showing greater interest in those he frequented (Maths room, Portuguese room, computer room and early intervention room).

At the end of the second meeting, it was explained that the model of the first floor had been built to a scale of 1:50. Again, the embossed line segment (1 cm) and the string (50 cm) were used for size perception.

In the third meeting, the previous activities were resumed, emphasising the definition and types of representation and the relationship of proportion. Scale was discussed again in relation to the two models. The aim was to reinforce the idea that scale applies to all elements of the representation:

> Researcher: The tip of your finger on the model (I place the student's finger on the top of the model) corresponds to how much in reality?
> Student B: Twenty-five.
> Researcher: That's right, because the scale is 1:25. So everything here is 1:25. A little piece on the table corresponds to how much on the real table?
> Student B: Twenty-five.
> Researcher: Very good. And on the shelf?
> Student B: Twenty-five.

This was followed by a comparative analysis of the models. As with the activity with student A, the following aspects were analysed: size of the space represented; total and partial size of the representations; number of places represented; level of detail and number and size of objects. Student B had difficulty understanding that the size of the space represented in the model of the first floor was larger than that of the Maths room, even though the size of the rooms in the first representation was smaller.

After resuming the discussion on scale, the activity began with different representations, which were first explored by the student. Similar to student A, student B carried out the proposal without difficulty and was able to define the adequacy/inadequacy of the representations in relation to the models, associating them with the different scales. In order to position the pilaster on the model of the first floor, the student asked where it was located in reality and asked for the text containing the number and name of the rooms (Figure 28), in order to include this element in the representation (Figure 29).

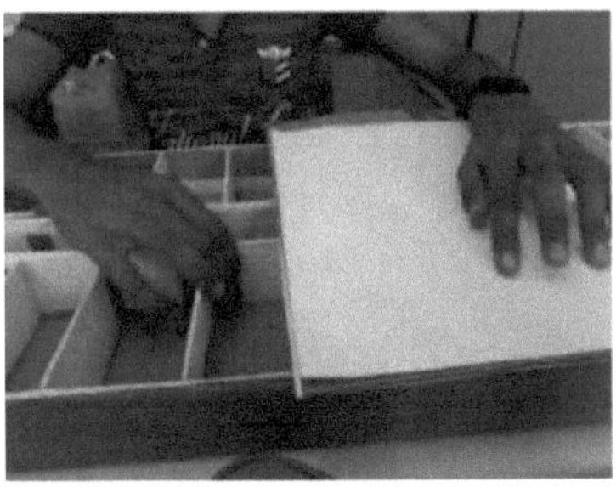

Figura 28 - Reading the explanatory text for analysing the first floor model, by student B
Photo by the author

Figura 29 - Positioning of the pilaster representation on the first floor model, by student B

Photo by the author

The student was then asked to reflect on the construction of a model of the entire CEBRAV building, on a scale of 1:100. Again, he was confused about the measurement (even when using the 100 cm string) and its influence on the representation. In this sense, it was decided to address only the size of the space that would be represented, or

i.e. the entire CEBRAV building. Asked about the size of the representation and the amount of detail, the student replied: "it depends on the scale".

This response showed that the student had understood proportional reasoning and scale, although these had not been consolidated quantitatively, inferring the need for more time to work on these concepts. Finally, the student stressed that he enjoyed the activity, above all because the models made it possible to understand the rooms, their spatial layout and internal organisation.

4.3 Student C

The activities with student C took place on 18 and 25 November 2014, in the afternoon, in the Maths workshop room. As in the meetings with the other two students, the activity began by explaining its organisation and investigating her previous knowledge of models, representation, proportion and scale. With regard to the first, the student said she had never worked on it and with regard to the other questions, although they had been covered at school, she was unable to verbalise any explanation.

The definition of representation was covered using the representations of the chair, the book and the lollipop and comparing them with their real-life counterparts. These resources also helped to understand the types of representation in relation to size (smaller, equal and larger) and to discuss the differentiation between representation and real objects.

When asked about the organisation of the Maths workshop room, the student said she knew very few elements, including tables, chairs and the materials used in class. It is interesting to problematise that student C classified the room as small, which makes it possible to infer that the number of elements known about a given space influences the perception of its size.

Although she defined the room as small, when she thought about the possible representation of this space on an A4 sheet of paper (perceived by touch), the student defended the need to reduce it. Considering the model as a reduced representation of reality, it was proposed that it be explored freely.

The limited understanding of the objects in the room made it difficult to explore the model, and it was necessary to anticipate the use of the legend in the interpretation (Figure 30). It should also be noted that this process was slow due to the difficulty of reading Braille and correlating the textures of the legend and the model. At the end of the exploration, the student was surprised by the number of objects in the room .[27]

[27] It was not possible to explore the real environment on the first day of the activity due to the number of students in the room.

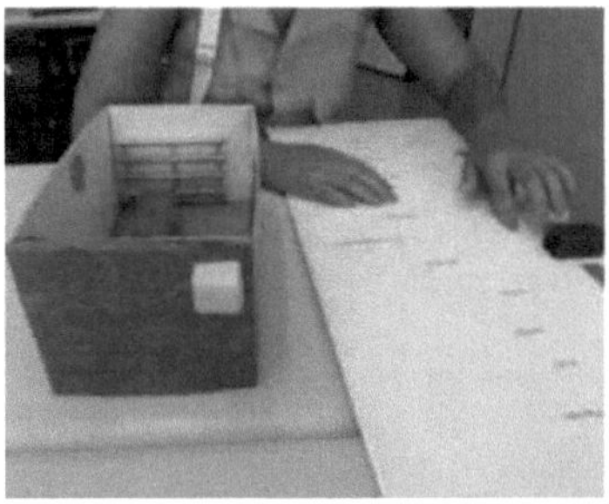

Figure 30 - Student C exploring the maths room model Photo by the author

Subsequently, the exploration of the model was guided by the perception of size and the proportional relationships established between the constituent parts of the elements, between these and the model as a whole, and between the model and reality. It was emphasised that if the model is smaller than the space represented, all its elements will be smaller, so that the proportional relationships are maintained.

The definition of scale was then explained, based on the perception of the 1 cm line segment and the 25 cm string. At this point, student C had difficulty understanding that the scale relationship applied to all the elements of the model.

The model of the first floor was shown, identifying the space represented. The student was asked to start the exploration by locating the corridor, but said that this representation was more complex due to its size. So the interpretation process was based on the legend, which made it possible to identify all the elements.

After exploring the model, she analysed the organisation of the rooms in the corridor, as well as their internal structure, which was made possible by the explanatory text (Figure 31). The student was surprised by the number of doors and rooms, pointing out that she only regularly knew the ones she went to (Maths room, Portuguese room, computer room and Braille room).

Figura 31 - Reading the explanatory text for analysing the first floor model, by student C
Photo by the author

Finally, she established that the model of the corridor was on a scale of 1:50, and the reasoning behind this correlated with the reasoning behind the model of the Maths workshop room. When exploring the 50 cm string, the student asked to compare it with the 25 cm string, concluding at the end: "it's much bigger". It is interesting to note that the measurements were not worked on from the perspective of the metric system, which the student had not mastered.

Asked if the size relationship defined by the scale applied to all the elements in the representation of the first floor, the student replied that it did, demonstrating: "a finger tip here (she positioned her finger at the top of the Maths room in the model of the first floor) corresponds to 50 pieces in the real thing".

In the second meeting, the discussions held on the first day of activities were resumed, and student C answered all the questions correctly, which made it possible to verify that she had understood the discussions.

Considering the 1:25 and 1:50 scales, it was decided that these influenced the characteristics of the representations. A comparative analysis of the two models was therefore proposed, based on the same questions that had been addressed with the other students: size of the space represented; total size and parts of the representation; number of places represented; level of detail and number and size of objects. The student understood the activity and answered all the questions correctly.

Prior to the activity with the various representations (chair, book, lollipop, table, cupboard and pillar), it was suggested to student C that she explore the Maths room in real life. In order to do this, she quickly recalled the elements represented in the model of this space (Figure 32), which were later confirmed in reality.

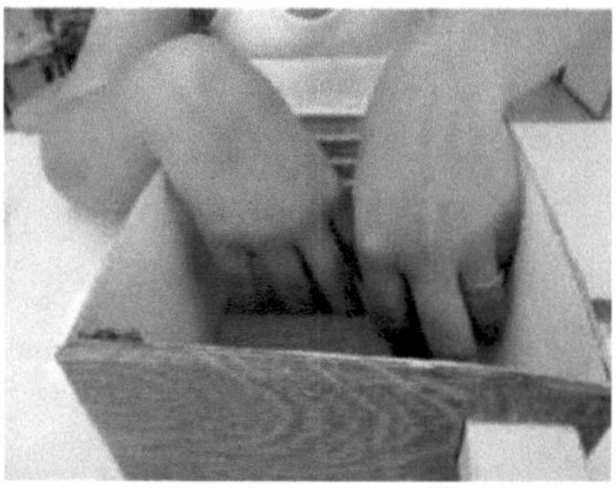

Figura 32 - Confirmation of the constituent elements of the maths room model by student C
Photo by the author

After exploring the real environment, the activity with the representations was carried out. Like the other two students, student C was able to define the adequacy/inadequacy of the representations in relation to the models, associating this context with the scales:

> Researcher: Where are the lollipops in the room?
> Student C: In the cupboard under the window.
> Researcher: Do we have a representation of a lollipop?
> Student C: Yes.
> Researcher: Then put that lollipop away in the cupboard.
> Student C: (Laughs) It won't fit.
> Researcher: Why do you think that?
> Student C: Because it's bigger than the cupboard.
> Researcher: If the lollipop is bigger than the cupboard, does that mean it's on the same or a different scale to the model?
> Student C: Different.

In the hypothetical activity of constructing a representation of the entire CEBRAV building, on a scale of 1:100, even though she had difficulties with the perception of the measure, the student came up with answers that showed understanding of proportional reasoning and scale.

Finally, in her assessment of the activity as a whole, similar to the other students, student C highlighted the contribution of the models to understanding spatial organisation, which in her case was considerably restricted.

4.4 Results and Discussions

The process of constructing the concept of scale, according to Le Sann (2007), goes beyond the proportional dimension and involves two other aspects: representation and comparison. In this sense, the approach taken to the definition of representation was considered adequate and enabled the students to understand its differentiation from the real object.

The supervising teacher praised the proposal to address the definition of representation, since it is widely used in teaching visually impaired students, as it makes it possible to access objects and elements that are often not possible to explore by touch in the real environment.

Another aspect of the definition of representation discussed was the types (smaller, equal and larger), defined according to size in relation to the object represented. This stage of the activity made it possible to discuss the second aspect presented by Le Sann (2007) in the construction of the concept of scale, comparison.

Thus, from the start of the activity, students were able to compare representations and their corresponding objects, which contributed to their understanding of the types of representation, as well as subsidising the construction of the notion of proportionality and the concept of scale.

It is also considered that the understanding of the types of representation was influenced by the use of everyday objects, given that even before exploring the real object and comparing it to the representation, the students were already aware of the size relationship established.

During the planning and execution stages of the didactic sequence, we sought to contribute to the participants' understanding of spatial organisation, as we understand that this influences the orientation and mobility skills of visually impaired people. This objective was achieved above all in relation to the Maths workshop room.

We found that exploring the model, combined with the perception of the room in real life, made it easier to identify the elements and their spatial arrangements. The analysis of the organisation of this space was based, among other aspects, on the topological relationships defined by Piaget, so it was decided to organise the observation and analysis process based on relationships such as: neighbourhood, separation and order. In addition to structuring the proposal, the enthusiasm of the students, especially student C, in carrying out this stage of the activity undoubtedly contributed to obtaining a positive and satisfactory result.

With regard to the first floor, unlike the Maths workshop room, the students' spatial perception was poor. The exploration of the model based on the explanatory text (room numbers and names) helped to improve this understanding. The students were very involved and interested, especially in relation to the rooms they attended at CEBRAV. It is recognised, however, that in order to consolidate and systematise their understanding of this space, they need more time than was available in the research to interpret and analyse the model, as well as to explore this environment in reality.

The construction of the notion of proportion, one of the main objectives of the pedagogical proposal, was approached based on the mathematical concept of proportional reasoning. Firstly, it was defined that the act of comparing two variables is intrinsic to the construction of the notion of proportionality.

Based on this assumption, the work with this notion was based on first and second order relationships (topic 2.2 of Chapter II), which, according to Spinillo (2002), are established between two or more sets. It is interesting to note that considering the cartographic scale, the proportion relationship applies to the size variable (distance) and therefore the first and second order relationships are based on this criterion.

In this work, however, we chose to consider other dimensions of the observed object in addition to the size variable, such as location, position in relation to other objects, materials and colours used in its representation and functionality.

Thus, by defining the model as a whole and the corresponding reality as another, in the first-order part/part relationship, we sought to analyse the representation of a given object in itself. This relationship was addressed during the exploration of the representation of the chair and some of the component elements of the Maths workshop room, which were analysed in isolation.

In the first-order, part/whole relationship, the object was analysed considering its particular characteristics and its insertion and relationship in the overall context of the model. For example, the analysis was directed towards the representation of the bookcase, so its location in the model of the room and its relationships (of size, distance, functionality, among others) with the other objects in this representation were considered and problematised.

The second-order relationship, in turn, was established by comparing the object or model with its real counterparts, since it is a relationship between sets. The notion of proportionality developed during the activity was based on non-quantifiable proportions, in which the size relationship is perceptual and not metric.

It is possible to say that larger scale models, which therefore represent smaller spaces, have a greater amount of detail and make it possible for the objects represented to resemble their real counterparts, are better suited to working with the notion of proportion, as they make it easier to establish first and second order relationships, as well as understanding spatial organisation, which influences the orientation and mobility skills of visually impaired students.

Finally, based on non-quantifiable proportionality, the concept of cartographic scale was worked on. Initially, the aim was to address quantifiable proportion, based on an understanding of the size relationship that was established between the representation and the real object/space. To do this, a 1 cm relief line segment and 25 cm, 50 cm and 100 cm strings were used.

It was clear that the students understood the reasoning behind the concept of scale, i.e. that one part in the model corresponded, for example, to 25 parts in reality and that this relationship applied to all the elements in the representation. However, students B and C had difficulties understanding the measurement itself.

Another aspect of the concept of scale that the students showed they understood was its influence on the characteristics of representations. At the end of the activity, they realised that the total and partial size of the representation and its level of detail are determined by the scale adopted, which results in different representations.

We believe that the work on calculating exact measurements (Euclidean relations), which had been planned, was not carried out. Firstly because of the fragile understanding of the metric system shown, above all, by students B and C, who had difficulty perceiving the sizes contained in the strings and their relationship with the size of the line segment.

In addition, during the course of the activity it was realised that the models used in this proposal (on large scales) make it difficult to move the fingers in the counting process, interfering with the definition of the representation's measure and, consequently, the mathematical calculation. In this sense, we believe that smaller scale models with flattened shapes, such as contour models, may be better suited to working with the calculation of measurements and, therefore, the quantifiable aspect of the concept of cartographic scale.

Given the relevance of this concept to the teaching of cartography and geography, it is proposed that working with larger scale models should be combined with other resources, such as smaller scale models, floor plans and tactile maps, which will make it possible to understand the non-quantifiable and quantifiable aspects of this concept, thus enabling visually impaired students to understand it in its entirety.

CHAPTER 5

FINAL CONSIDERATIONS

The concept of inclusive education, by establishing the regular school system as the locus of education for students with disabilities, makes it essential and urgent to reflect on the educational specificities of these subjects. In the state of Goiás, inclusive education has been structured on the basis of legislation and guidelines, although practice and everyday school life show that inclusion policies have not yet really materialised.

One of the factors needed to change this context is teaching work, which in turn is related to the teacher training process. Universities, however, have not satisfactorily addressed the issues of inclusion, diversity and teaching students with disabilities in undergraduate programmes, making it difficult to overcome the obstacles to inclusive education.

It is recognised that different types of disability present different demands, which require the adoption of specific knowledge, practices and resources, and it is unlikely that initial training will take into account all the specificities involved in teaching students with disabilities. However, ignoring this context is definitely not an option for teacher training courses.

When they leave university, newly qualified teachers won't have all the knowledge they need to work with students with sensory, motor and/or intellectual disabilities. But when they graduate, they need to be aware of the fundamentals of inclusion, the presence of students with and without disabilities in the classroom and the dynamics and challenges that this reality poses. What's more, initial training should provide possibilities and alternatives for those who want to specialise in these issues.

This problem between teacher training and work in Inclusive Education includes the teaching of students with visual impairments, for whom specific legislation and planning is intended, but which in reality continues to be led by pedagogical practices that are closer to exclusion than inclusion.

In Goiás, in recent years, partnerships have been established between university institutes, schools and the specialised care centre (CEBRAV), which undoubtedly contributes to the initial training of undergraduates and the practice of professionals in these institutions, but does not solve the problem.

School Geography stands out in this context, as it contributes to citizen education by enabling students to analyse and critically understand space. Cartography is one of the essential skills for building spatial reasoning, as it makes it possible to represent, analyse and understand phenomena and objects through the lens of spatiality.

Based on the assumption that Cartography is established as a language based on visual graphic variables and that the analyses carried out here are aimed at visually impaired students, the accessibility of cartographic knowledge to DV subjects is made possible by the products and practices of Tactile Cartography.

This science (art and technique) is established as a branch of Cartography that has specific theoretical and methodological foundations, and is responsible for designing and making cartographic products based on the principles of tactile graphic language. The diversity of materials and techniques available in Tactile Cartography expands the possibilities for its use in different environments and contexts.

Tactile cartographic products are generally constructed by combining tactile and visual information. Thus, working with these resources is not restricted to visually impaired individuals, but is suitable for all students, which contributes to the consolidation of a more inclusive education that is in line with the educational context that is currently being structured .

It should be noted that the simple use of suitable materials does not guarantee student learning, i.e. the use of a tactile map will not make it possible for DV students to learn cartographic/geographic knowledge if it is dissociated from the teacher's mediation. It is therefore essential that, in addition to resources, teaching methodologies and practices are adopted that are adapted to the specific needs of the students, of all students.

One of the most important concepts in cartography is the cartographic scale, which affects both the process of building and analysing cartographic products. Working with this concept is seen as one of the main challenges in teaching cartography, which is mainly due to its proximity to mathematical reasoning.

The difficulty attributed to understanding the cartographic scale stems mainly from its restriction to the quantifiable proportional notion, i.e. this concept is approached solely by algorithmising and calculating measurements. However, the notion of proportion is first established in non-quantifiable terms, which must be taken into account when teaching this concept.

The relationship of proportion, when considered in cartographic scale, focuses on the distance (size) variable, which is established between the object/space represented and its representation. Approaching the notion of this variable primarily from the point of view of perceptual measures and non-quantifiable judgements facilitates its understanding and, consequently, the construction of the concept of cartographic scale. In this way, observing and analysing the places where people live becomes an opportunity to put this knowledge into practice, which in turn becomes more meaningful to the students.

In the case of visually impaired students, working with non-quantifiable proportionality is even more necessary due to their difficulty in perceiving measurements, which in the concept of scale is established between extremely small sizes in the representation and larger sizes in the corresponding real space.

Tactile models, especially at larger scales, are very suitable for understanding non-quantifiable proportional notions, since they represent spaces close to the students (classrooms, bedrooms, etc.) and make it easier to understand first-order (part-part and part-whole) and second-order relationships.

In addition, the size relationships established in these representations relate to closer measurements, for example, 1 cm in the model corresponds to 10 cm, 20 cm, 30 cm in reality, and it is possible to make it easier for students to perceive size using string, embossed line segments, etc. Analyses of the activities carried out showed that the research subjects understood proportional reasoning and scale in non-quantifiable terms.

Working with the calculation of measurements has not been properly explored, which is a possibility for future studies and research. It is suggested that this topic (calculating measurements) be carried out with models on a smaller scale and with flattened shapes, which will allow for better finger movement, perception of the measurements on the model and calculations. Still on the quantifiable aspect, considering the relationship between discrepant sizes, such as 1:10,000, 1:100,000, it can be deduced that floor plans and tactile maps are also alternatives.

In this way, the use of larger scale models combined with other resources, such as smaller scale models and two-dimensional representations, is a proposal for working with the concept of scale, as it approaches it from its non-quantifiable and quantifiable aspects, overcoming the limitation of this to algorithmisation.

Finally, it is essential to point out that when working with tactile cartographic products that represent students' lived spaces, regardless of the proposal, the concept or the content that is to be addressed, the spatial organisation of the places represented must be problematised at some point. This aspect became evident during the activity proposed here, and was confirmed in the evaluation stage, when all the research subjects pointed out that the importance of the models was mainly because they made it possible to understand spatial organisation.

It is known that understanding the materiality of space, the elements that make it up and their arrangements is not enough to analyse geography, which involves other spatial dimensions. However, when these issues are problematised with visually impaired students, it expands their possibilities of understanding the place where they live, which in turn directly influences the improvement of their orientation and mobility skills.

Therefore, since Geography is the science and school subject responsible for enabling the critical analysis of space, it is its duty to create conditions for visually impaired students to understand it on various scales, among which the local scale, as an immediate and lived space, with its materiality and specific organisation, becomes essential.

BIBLIOGRAPHICAL REFERENCES

AGUIAR, M. C. A. de. The formation of fraction and proportionality concepts based on Piagetian theory. **Psicologia:** Ciência e Profissão. Brasília, v.3, n. 2, 1983. Available at: www.scielo.br/scielo

ALMELDA, D. B. de. **From special to inclusive?** : a study of the school inclusion proposal of the Goiás State Network, in the municipality of Goiânia. PhD Thesis, Faculty of Education, State University of Campinas, Campinas, 2003.

ALMEIDA, L. C. de; NOGUEIRA, R. E. Cartographic initiation of blind adults. In: NOGUEIRA, R. E. **Motivações hodiernas para ensinar Geografia**: representações do espaço para visuais e blinduais. Florianópolis, 2009.

ALMEIDA, R. A. de. Tactile Cartography at USP: two decades of research and teaching. In: Book: Tactile Cartography: orientation and mobility for the visually impaired. In: FREITAS, M. I. C. de; VENTORINI, S. E. **Cartografia tátil**: orientação e mobilidade às pessoas com deficiência visual. Jundiaí, Paco Editorial: 2011.

ALMEIDA, R. D. de; PASSINI, E. Y. **O Espaço Geográfico**: Ensino e Representação. São Paulo: Contexto, 6 ed, 1998.

BATISTA, C. G. Concept Formation in Blind Children: Theoretical Issues and Educational Implications. **Psicologia**: Teoria e Pesquisa, Campinas, v. 21, n. 1, p. 7-15, Jan-Apr, 2005. Available at: http://www.scielo.br/pdf

BEZERRA, G. F.; ARAUJO, D. A. de C. New (re)configurations in the Ministry of Education: between Ariadne's thread and Penelope's shroud. **Revista Brasileira de Educação**. v. 19, n. 56, jan-mar, 2014. Available at: http://www.scielo.br/scielo

BRANDÃO, C. R. (org.) **Participatory Research**. São Paulo: Brasiliense, 1999.

BRAZIL. Ministry of Education. Special Education Secretariat. **National Guidelines for Special Education in Basic Education**. Special Education Secretariat - MEC/SEESP, 2001. Available at: http://portal.mec.gov.br/seesp

________ . Ministry of Education. Department of Special Education. **Knowledge and practices of inclusion**: Developing competences for meeting the special educational needs of blind and low-vision students. Brasilia, 2006.

________ . Ministry of Education. Law n. 9394/96. **Lei de Diretrizes e Bases da Educação Nacional**, Brasília, 5. ed, 2010. Available at: http://bd.camara.gov.br

. Ministry of Education. Department of Special Education. **DECREE NO. 7.611.** Nov, 2011. Available at: http://www.planalto.gov.br/ccivil

BRUNO, M. M. G. **O desenvolvimento integral do portador de deficiência visual**: da intervenção precoce à integração escolar. São Paulo: NEWSWORK, 1993.

BUENO, M. A. **Municipal school atlases and the possibility of continuing teacher training**: a study of case in Sena Madureira/AC. PhD Thesis, Institute of Geosciences, UNICAMP, Campinas, 2008.

BUTTIMER, A. Learning the dynamism of the lived world. Transcribed from Annals of the Association of American Geographers, 66 (2): 277-292, Jul, 1976. Translated by Neide Piran and Antônio Christofolletti. In: CHRISTOFOLLETTI, A. (org.). **Perspectivas da Geografia**. Rio Claro, SP: DIFEL, 1982.

CARMO, W. R. do. **Cartografia Tátil Escolar**: experiências com a construção de materiais didáticos e com a formação continuada de professores. São Paulo, 2009. Originally presented as a master's thesis, University of São Paulo, 2009.

________ . Teacher training in tactile cartography: theoretical issues and practical experiences. In: FREITAS, M. I. C.

de; VENTORINI, S. E. **Cartografia Tátil**: orientação e mobilidade às pessoas com deficiência visual. Jundiaí: Paco Editorial, 2011.

CARVALHO, L. R.; CARVALHO, S. D. de; CUSTÓDIO, W. G. (org.) **Specialised Educational Assistance - AEE**: perspectives for inclusive education in Goiás. Notebook 2. Coordination of Special Education, 2010.

CASTELLAR, S. M. V. **Geographical Education**: Psychogenetics and school knowledge. **Cad. Cedes**, Campinas, v. 25, n. 66, p. 209-225, May/Aug. 2005. Available at: www.cedes.unicamp.br

CAVALCANTI, L. de S. Theoretical conceptions and elements of Geography teaching practice. In: CAVALCANTI, L. de S. **Geografia e práticas de ensino**. Goiânia: Editora Alternativa, 2002.

_______ . **Geography teaching at school**. Campinas, SP: Papirus, 2012.

BRAZILIAN CENTRE FOR REHABILITATION AND SUPPORT FOR THE VISUALLY IMPAIRED. State Department of Education, Goiás, 2011. Available at: http://www.cebravcap.go.gov.br

CMDV. **Site Portal do Deficiente Visual**, 2014. Available at: www.cmdv.com.br

CORRÊA, R. L. Space: a key concept in Geography. In: CASTRO, I. E. de; GOMES, P. C. da C.; CORRÊA, R. L. **Geografia**: Conceitos e Temas. Rio de Janeiro: Bertrand Brasil, 11ª ed., 2008.

DARDEL, E. **O homem e a terra**: natureza da realidade geográfica; translation: HOLZER, W. São Paulo: Perspectiva, 2011.

DEMO, P. **Participatory research**: myth and reality. Brasília: UNB/INEP, 1982. Available at: http://www.dominiopublico.gov.br/download/texto/me001993.pdf

DUARTE, P. A. **Scale**: fundamentals. Florianópolis: Ed. da UFSC, ed. 2, 1983.

FONSECA, V da. **Special education**. Porto Alegre: Artes Médicas, 1987.

FREITAS, M. I. C. de; VENTORINI, S. E. **Tactile cartography**: orientation and mobility for the visually impaired. Jundiaí, Paco Editorial: 2011.

GLAT, R.; FERNANDES, E. M. Da Educação Segregada à Educação Inclusiva: uma Breve Reflexão sobre os Paradigmas Educacionais no Contexto da Educação especial Brasileira.

Revista Inclusão, n. 1, p. 1- 6, MEC/ SEESP, 2005. Available at: http://www.eduinclusivapesq-uerj.pro.br

GOIÁS. State Department of Education. Special Education Department. **Operational Guidelines of the Special Education Department for the Inclusion Support Network**. n.d.

_______ . STATE EDUCATION COUNCIL. **CEE Resolution no. 07**. December 2006. Available at: http://www.cee.go.gov.br/?p=841

. State Department of Education. **Curriculum reorientation for 1st to 9th grade**: curriculum in debate - Goiás: didactic sequence: invitation to action: geography: notebook 6.6. Goiânia: poligráfica, 2009.

_______ . State Department of Education. **Reference Curriculum for the Goiás State Education Network**. Goiânia, 2012.

HAGUETTE, T. M. F. **Metodologias Qualitativas na Sociologia**. Petrópolis/RJ: Editora Vozes, 8 ed., 2001.

HARVEY, D. O Espaço como Palavra-chave, 2006 (Translation: Letícia Gianella). In: HAESBAERT, R; NUNES, J. **Technical review**, 2010.

HOLZER, W. . Humanist Geography: a review. **SPACE AND CULTURE**, Commemorative Edition, Rio de Janeiro, 1993-2008.

_______ . A phenomenological discussion on the concepts of Landscape and Place, Territory and Environment. **Revista TERRITÓRIO**, ano II, n. 3, jul-dez, 1997. Available at: arquivo.rosana.unesp.br/docentes

ITABORAHY, N. Z. **Uma reflexão sobre a Pesquisa Participante em Geografia**: lugares em construção. s/d. Available at: http://www.ufjf.br

JORDÃO, B. G. F. **Tactile cartography for visually impaired students: the experience of the adapted globe.** Monograph, Universidade Estadual Paulista, Ourinhos, 2011. Available at: www.bv.fapesp.br/pt/bolsas/126783/cartografia-tatil-a-experiencia-do-globo-adaptado

TACTILE AND SCHOOL CARTOGRAPHY LABORATORY. Federal University of Santa Catarina, Department of Geosciences, Santa Catarina, 2011. Available at: http://www.labtate.ufsc.br/

LE SANN, J. G. The notion of scale in Cartography. **Geography and Teaching Magazine**. Belo Horizonte - MG, year II, n. 5, 1984

. **Trails for Geography**: space and time 5th grade. Belo Horizonte: Dimensão, 2007

LOWENTHAL, D. **How we know the past**. São Paulo: Projeto História, 1998.

MAKHOUL, C. S.; et. al. (org). **State programme for diversity education from an inclusive perspective**: 10 years. Notebook 9. Coordination of Special Education, 2010.

MATINELLI, M. **Cartografia Temática**: Caderno de Mapas. São Paulo: Editora da Universidade de São Paulo, 2003a.

________ . **Maps of geography and thematic cartography**. São Paulo: context, 2003b.

MASI, I. de; et. al. Ministry of Education. Special Education Secretariat. **National Programme to Support the Education of the Visually Impaired:** Teacher Training. Brasilia, 2002

MASINI, E. F. S. A educação do portador de deficiência visual: as perspectivas do vidente e não vidente. **Em Aberto**, Brasília, year 13, n. 60, Oct-Dec, 1993. Available at: http://www.rbep.inep.gov.br/index.php

MAZZOTTA, M. J. S. **Trabalho docente e formação de professores de educação especial**. São Paulo: EPU, 1993.

________ . **Special education in Brazil**: history and public policies. São Paulo: Cortez, 1996.

________ . **Special education in Brazil:** history and public policies. 5. ed. São Paulo: Cortez, 2005.

MIRANDA, S. R. da S.; PAULA, M. M. de. Specialised Educational Assistance: visual impairment. In: CARVALHO, L. R.; CARVALHO, S. D. de; CUSTÓDIO, W. G. (org.). **Specialised Educational Assistance - AEE**: perspectives for inclusive education in Goiás. Notebook 2. Coordination of Special Education, 2010.

MONTOITO, R.; LEIVAS, J. C. P. The representation of space in children, according to Piaget: the mental processes that lead to the formation of the notion of Euclidean space. **VIDYA**, v. 32, n. 2, p.21-35, jul./dez., 2012.

PACHECO, L. B.; et. al (org.). **Inclusive Education and Specialised Services**: CAP, CAS, NAAH/S and Project Today. Notebook 7. Coordination of Special Education, 2010.

PAGANELLI, T. I.; ANTUNES, A. do R.; SOIHET, R. **The notion of space and time**: the map and the graph. Rio de Janeiro, 1981.

PASSINI, E. Y.; ALMEIDA, R. D. de; MARTINELLI, M. Cartography for Children: literacy, education or cartographic initiation. **Boletim de Geografia**. n. 17, 1999, p. 125135.

PIAGET, J. **The construction of reality in the child** (Translation: Álvaro Cabral). Rio de Janeiro: Zahar Editores. 3 ed., 1979.

PIAGET, J.; INHELDER, B. **The representation of space in children**. Porto Alegre: Artes médicas, 1993.

PISSINATI, M. C.; ARCHELA, R. S. Fundamentals of cartographic literacy in geography teaching. Geografia, v. 16, n. 1, jan./jun. 2007.

RODRIGUES, D. Ten (mis)ideas about Inclusive Education. In: RODRIGUES, D. (org.) **Inclusão e Educação:** doze olhares sobre a Educação Inclusiva. São Paulo: Summus, 2006.

RUIZ, A. R.; CARVALHO, M. P. de. The concept of proportionality. **Faculty of Education**. São Paulo, 131, Jan./Dec., 1980.

SANTOS, M. P. dos. Inclusive education: redefining special education. **Ponto de Vista**, Florianópolis, n. 3/4, 2002. Available at: http://www.perspectiva.ufsc.br/pontodevista

SASSAKI, R. K. TERMINOLOGIA SOBRE DEFICIÊNCIA NA ERA DA INCLUSÃO. **Educação online**. 6 ed., September 2005. Available: at:www.educacaoonline.pro.br

SCHMID, C. **Henri Lefebvre's theory of the production of space: towards a three-dimensional dialectic** (Translation: Marta Inez Medeiros Marques; Marcelo Barreto). **GEOUSP** - espaço e tempo, São Paulo, n.32, p. 89-109, 2012.

SEEMANN, J. Teaching Cartography that is not in the curriculum: Cartographic Views, "Carto-Facts" and "Cartographic Culture". NUNES. **Teaching geography**: new perspectives and practices. Dourados, MS : UFGD, 2011.
________ . **Carto-crônicas:** a journey through the world of cartography. Fortaleza: Expressão gráfica e Editora, 2013.

SENA, C. C. R. G. de. **The Study of the Environment as an Instrument for Teaching Geography**: Unveiling Pico do Jaraguá for the Visually Impaired. Master's dissertation, Department of Geography, USP, São Paulo, 2001.
________ . **Tactile cartography in the teaching of Geography**: a methodological proposal for the development and association of didactic resources adapted to visually impaired people. PhD Thesis, Department of Geography, USP, São Paulo, 2008.

SILVA, F. G. D. **The invisible forms of lived space**: understanding the CEBRAV corridor through a model and tactile floor plan. Goiânia, 2011. Originally presented as a monograph, Federal University of Goiás, 2011.

SILVA, M. A. B. da. **Municipal school atlases as mediators in the process of constructing geographical concepts:** a journey between everyday knowledge and scientific knowledge. undated.

SIMIELLI, M. E. R. Cartography in primary and secondary education. In: CARLOS, Ana Fani Alessandri (org). **Geography in the classroom**. São Paulo: Contexto, 1999.

SPINILLO, A. G. Proportional reasoning in children: considerations about educational alternatives. **Pro-Posições**. v. 5, n. 1 [13], Mar. 1994. Available at: www.proposicoes.fe.unicamp.br
________ . The role of specific interventions in children's understanding of proportion. **Psychology**: Reflection and Criticism. Porto Alegre, v.15, n.3, 2002. Available at: www.scielo.br/scielo

TARTUCI, D.; CARDOSO, C. R.; FREITAS, A. de O. Special education services in Goiás: what the Inclusion Policy Guidelines say about the attributions of its teachers. **Proceedings of the 1st International Symposium on Disability Studies**. São Paulo, June 2013. Available at: http://www.memorialdainclusao.sp.gov.br/br

TUAN, Yi-Fi. **Topophilia**: A Study of Environmental Perception, Attitudes and Values. Translated by Lívia de Oliveira. São Paulo: DIFEL, 1980.
__________________ . **Space and Place:** the perspective of experience. Translated by Lívia de Oliveira. São Paulo: DIFEL, 1983.

UNESCO. **Salamanca Declaration and line of action on special educational needs**, Brasilia: CORDE, 1994. Available at: http: _____ . **Convention on the Rights of Persons with Disabilities:** Optional Protocol to the Convention on the Rights of Persons with Disabilities. Official translation Brazil. Brasilia, 2007.

VASCONCELLOS, R. **Tactile Cartography and the Visually Impaired**: an evaluation of the stages of map production and use. PhD Thesis, Department of Geography, USP. São Paulo, 1993.

VENTORINI, S. E. **Experience as a determining factor in the spatial representation of the visually impaired**. Master's dissertation, Institute of Geosciences and Exact Sciences, UNIVERSIDADE ESTADUAL PAULISTA, Rio Claro, 2007.
_______________ . **Graphic representation and tactile cartographic language**: case studies. PhD Thesis,

Institute of Geosciences and Exact Sciences, UNIVERSIDADE ESTADUAL PAULISTA, Rio Claro, 2012.

ZABALA, A. A **prática educativa**: como ensinar. Porto Alegre: Editora Artes Médicas Sul Ltda., 1998. Available at: demogidascruzes.edunet.sp.gov.br

ANNEXES

Annex 1:

Aluminium map construction	
Materials needed:	Procedures:
• Aluminium blade with a thickness of 0.10 millimetres sold in hardware shops; • Rubber sheet or E.V.A., which will be the soft surface to make it easier to mark each sign on the blade; • Linear tools (reels of various sizes), point tools (tweezers, dry points, pen tips, pencil sharpeners) and zonal tools (textured boards, sandpaper, wood); • Spatula or wooden toothpick and needle for finishing.	0 map is drawn on tracing paper so that it can be transposed onto aluminium on the reverse side, as the entire drawing must be worked on the back of the foil to obtain the expected result. Always working backwards, mark out the lines, points and areas you want to represent with reels, dry points and other tools. It is constantly necessary to turn the blade and reinforce the markings on the reverse side with a wooden spatula, which helps to define the lines.

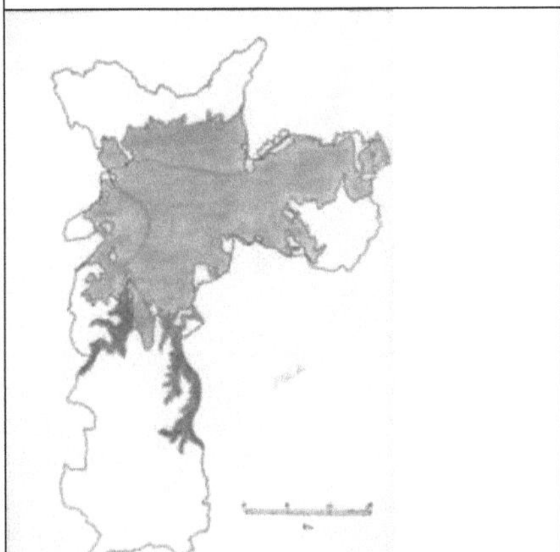

1. Draw the map on tracing paper.

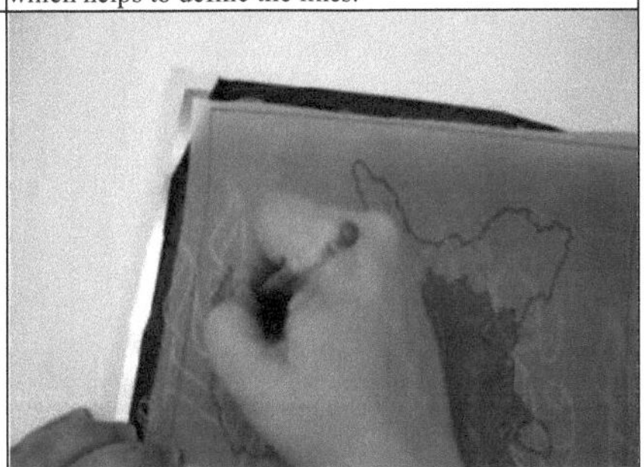

2. Copy the map inside out onto the aluminium foil.

3. Mark the copied design on the aluminium with a spool or dry point.

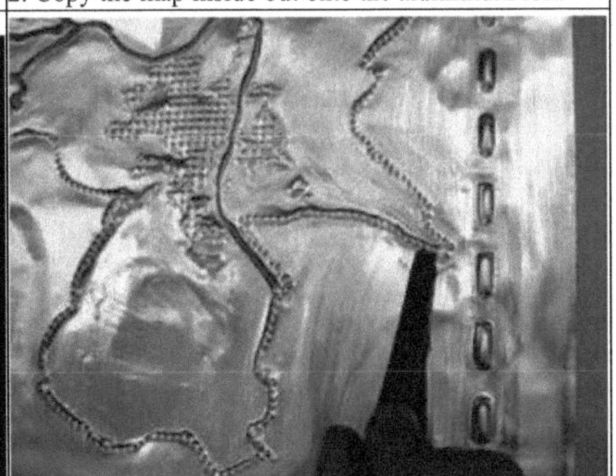

4. Use a spatula or wooden toothpick to define the lines on the aluminium.

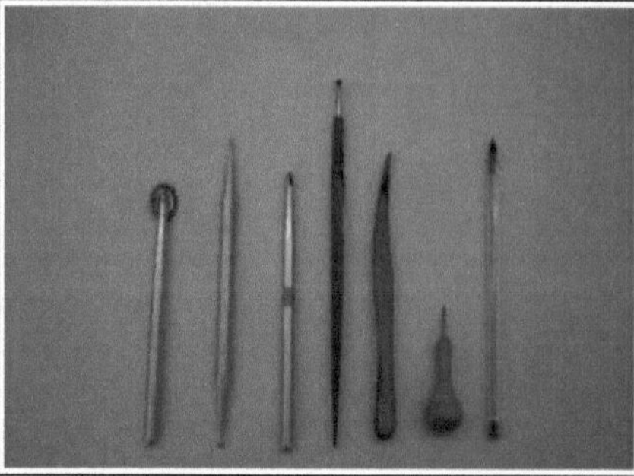

5. Lines can be marked with spools, dry points, pens or the Braille writing punch.

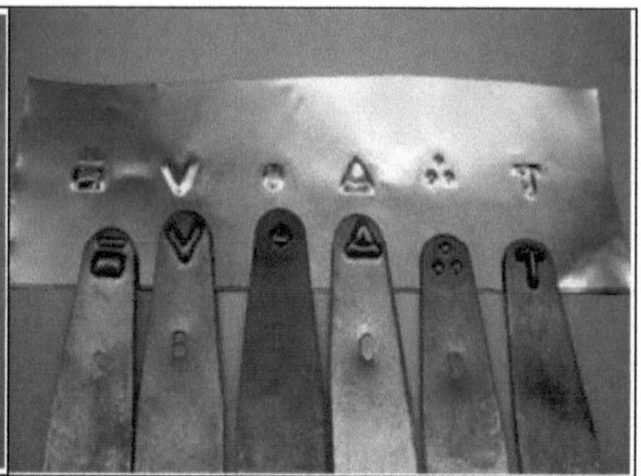

6. For point symbols, strike surfaces with a rubberised stick that have geometric shapes.

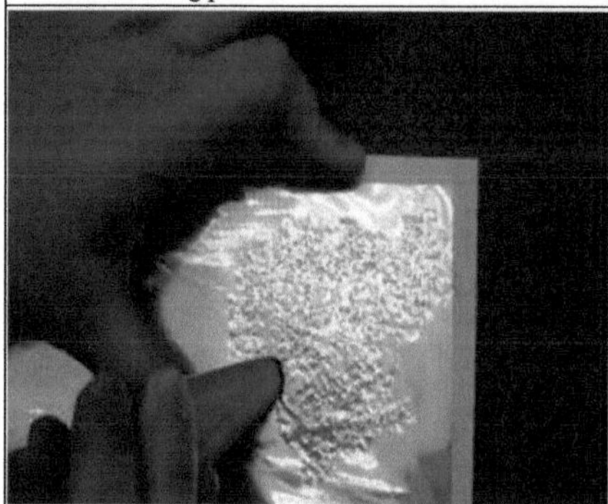

7. Filling in textured areas can be achieved by marking the aluminium with a stick on a textured surface (sandpaper, wood or metal).

8. Tactile representations made in aluminium can be completed with Braille and, if they are used as matrices for plastic copies, it is important to make small holes in the foil to help define the representation on the plastic.

Source: Sena (2008)

Annex 2:

Building maps with collage	
Materials needed:	Procedures:
• Cardboard, paperboard or cardboard to serve as a base for the representation.	The map is drawn on tracing paper and transposed with carbon paper onto the base.
• Textured surfaces: fabrics, sandpaper, cork, cardboard, etc.	The reverse side of the map is used to mark the sandpaper or cardboard used to fill in areas on the map with carbon.
• Twine, embroidery thread, string, mock-up sticks.	
• Buttons, beads and small objects that can represent specific symbols.	The selected lines, dots and textures are gently glued onto the map design with white glue.
• White glue and paintbrush.	

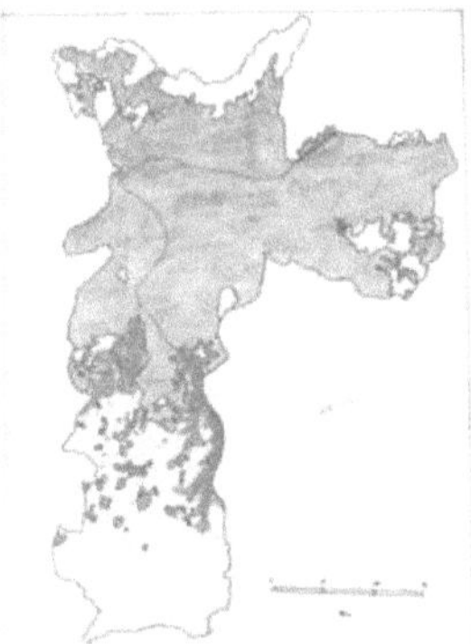	
1. Draw the map on tracing paper.	2. Transpose the map onto the sandpaper.
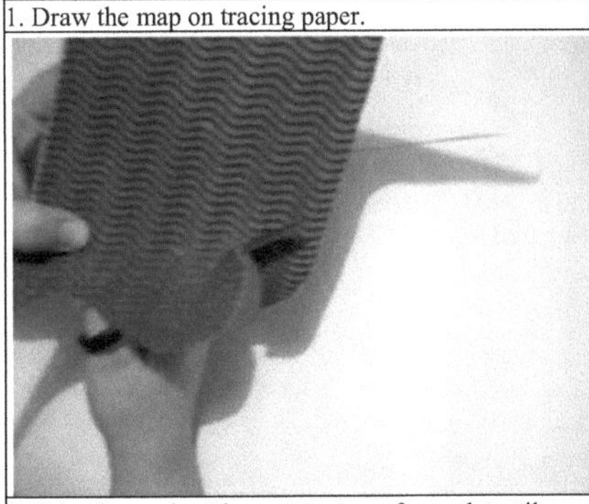	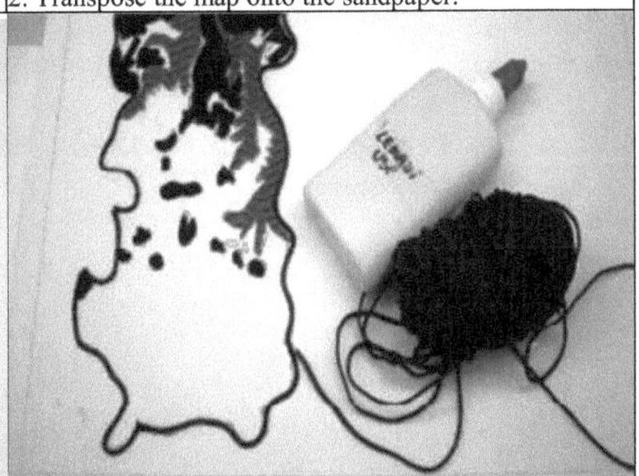
3. Selecting and cutting out textures for each tactile variable	4. Gluing the string to delimit the map
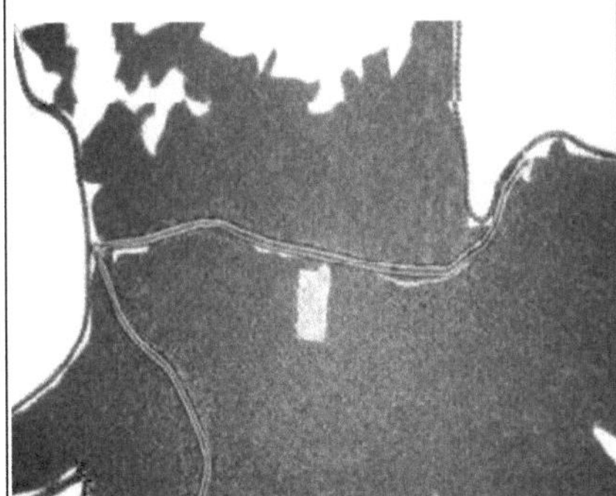	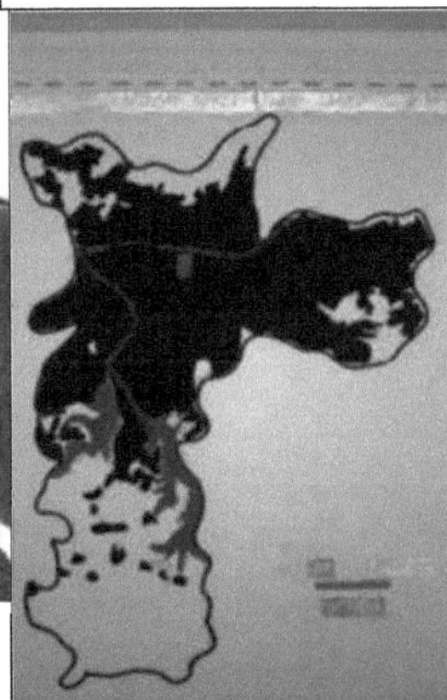
5. Composition of textures: in the map above the sandpapers have been superimposed to differentiate the urban area of São Paulo at different times.	6. The title and scale were identified with aluminium braille.

Source: Sena (2008)

Annex 3:

Building tactile graphic representations with cold porcelain	
Materials needed:	Procedures:
• Cold porcelain in different colours;	Draw the map on the base and outline the boundaries with

• High-relief fabric paint; • White glue; • Basis for drawing up the model (varies according to the type of representation).	embossed fabric paint. Modelling small amounts of cold porcelain, filling in the spaces delimited by the paint. In the case of representations with the variable height, overlay successive layers of cold porcelain, modelling the desired shape. Cover the entire representation with white glue
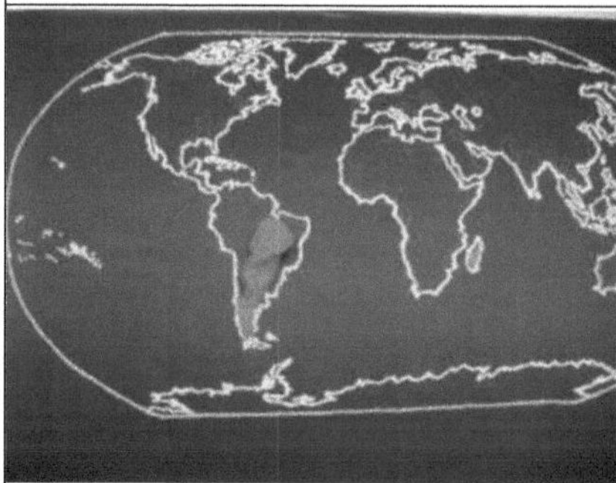	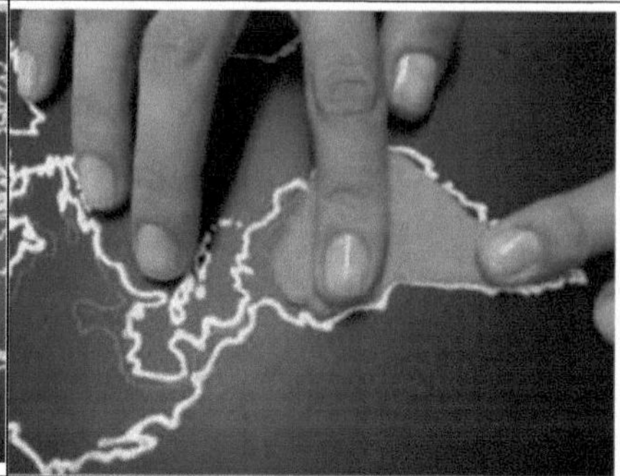
1. mark the area with raised paint.	2. Fill in the marked areas with successive layers of cold porcelain, modelling the dough until it reaches the desired elevation.
	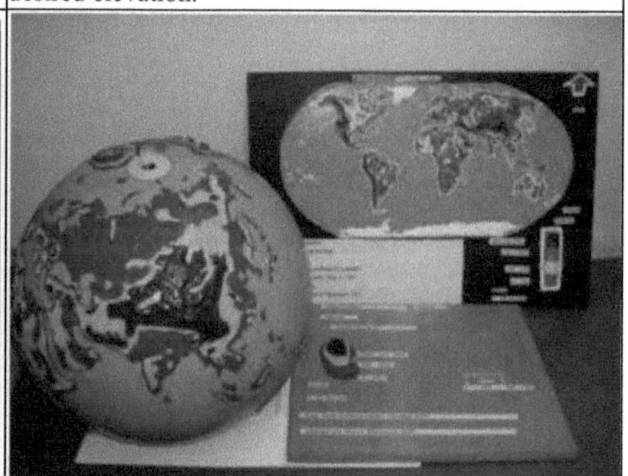
3. Use different colours of cold porcelain to reinforce the themes represented.	4. Cover the entire model with white glue to increase its resistance.

Source: Sena (2008)

Annex 4:

Screen printing of maps	
Materials needed:	Procedures:
• Tracing paper, transparency or photolithography. • Light table. • Special emulsion for developing. • Wooden frame with nylon mesh • Heat-expanding paint (puff) • Squeegee • Cardboard, paper or fabric.	0 map is drawn on tracing paper or plastic (transparency). 0 drawing of the map is engraved on the nylon canvas using photographic emulsion and a light table. Once the canvas is ready, the ink is printed onto the cardboard paper by running the squeegee over the canvas with the ink, spreading it out. After the paint has dried, invert the cardboard and iron it with a hot iron, subjecting the paint to the heat so that it takes on an embossed look.

1. maps printed on nylon canvas. For each colour used in the printing, a screen has to be constructed.

2. Fix the base to be printed on to the screen and distribute the ink around the edge of the screen.

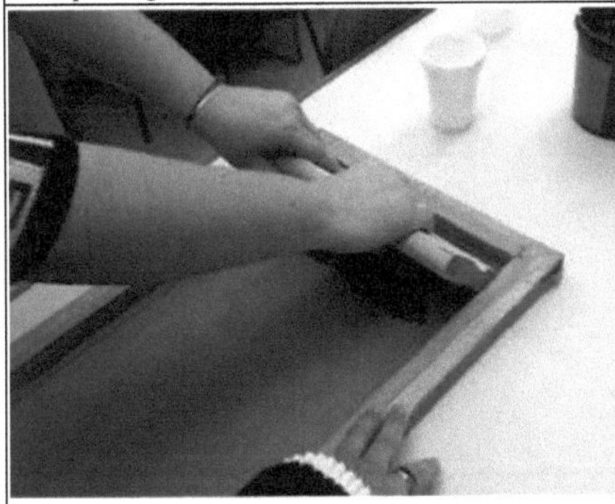

1

3. Spread the ink evenly over the area to be printed with the squeegee

4. Use the squeegee again to distribute the paint over the canvas.

5. If the map has more than one colour, wait for the first colour to dry before attaching the second colour.

6. Repeat the paint distribution procedure with the squeegee.

	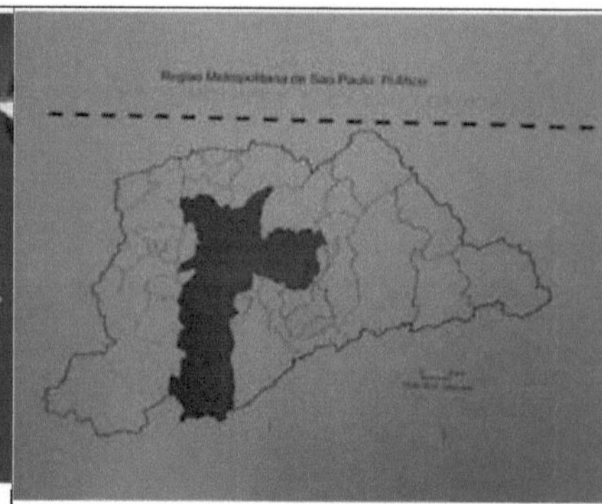
7. With the paint dry, invert the base and iron with a hot iron to make the paint expand.	8. Once the ink has expanded, the map can be used.

Source: Sena (2008)

Annex 5:

Map printed on microcapsulated paper being placed in the heating equipment
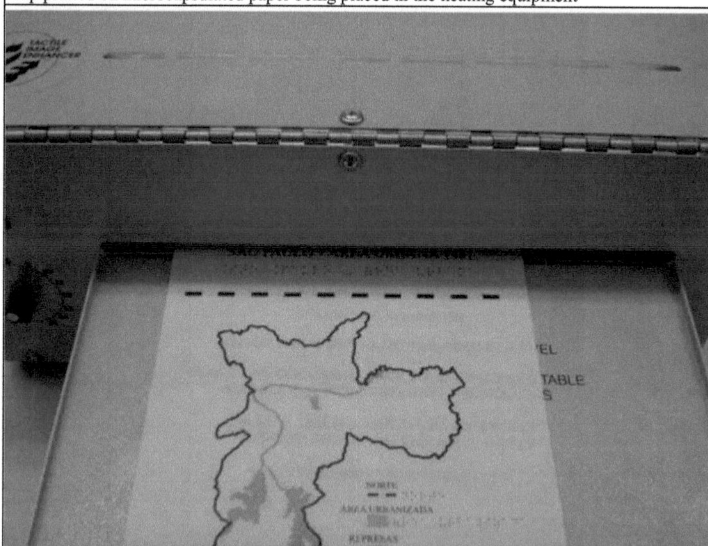
Paper gaining texture in the black printed areas

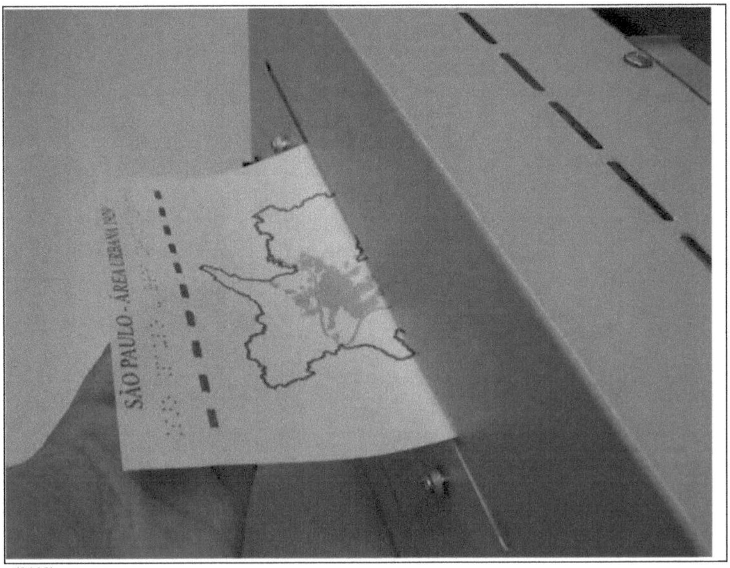

Source: Sena (2008)

Annex 6:

Observation guide

Field of research: Brazilian Centre for Rehabilitation and Support of the Visually Impaired - CEBRAV

Service:

Teacher:

Regarding the teacher's attendance and performance:

1) Theme/content

2) Methodology

3) Didactic-pedagogical resources used

4) Organisation of the room (light control, seating arrangements, etc.)

As for the students:

1) Educational specificity (what materials and environmental conditions they need)

2) Difficulty with the topic/content covered

3) Understanding of the proposed activities

4) Level of involvement with other students and the teacher

5) Orientation and mobility skills

Annex 7:

Door order

1 - Maths workshop

2 - Portuguese Workshop

3 - Pedagogical Coordination

4 - Social Services

5 - IT

6 - Braille Literacy

7 - Lift

8 - Production

9 - Early Intervention

10 - Psychology

11 - Ophthalmological Clinic

12 - Visual Rehabilitation

MIX
Papier aus verantwortungsvollen Quellen
Paper from responsible sources
FSC® C105338

Printed by Books on Demand GmbH, Norderstedt / Germany